MICROEXPLORERS

First edition for the United States and Canada
published exclusively 1997 by Barron's Educational Series, Inc.

Originally published in English under the title
How The Y Makes The Guy © Copyright Useful Books, S.L.,
1997 Barcelona, Spain.

Authors: Norbert Landa and Patrick A. Baeuerle
Illustrators: Antonio Muñoz, Ali Garousi and Roser Rius
Graphic Design: IGS - Barcelona, Spain
Illustration on page 37 according to Stilling-Hertel.

Address all inquiries to:
Barron's Educational Series, Inc.
250 Wireless Boulevard
Hauppauge, NY 11788

Library of Congress Catalog Card No. 97-74824

International Standard Book Number 0-7641-5064-2

Printed in Spain

9 8 7 6 5 4 3 2 1

MICROEXPLORERS

*A guided tour through
the marvels of
inheritance
and growth*

*Patrick A. Baeuerle
and Norbert Landa*

Welcome!

When I was a child, I wanted so much to know what I would look like when I grew up. Well, now I know. I have inherited my mother's nose, my father's eyes and hair, and my grandfather's necktie.

Seriously though, who do you look like? You probably look like your mother in some ways and like your father in others. Maybe you even look like your long-deceased great-grandmother. In any case, we all look more like people we are related to than like our best friends.

Compared with other forms of life, for instance a dog or an apple tree, people are all pretty much similar. Somewhere in every living being are plans that guarantee that dogs look like and act like dogs and people like people—although each individual looks or acts a little bit different.

What kinds of plans exist, and where do we get them? Are there plans for long noses and blue eyes? Does a set of twins have the same plans and all other people different ones?

We call such plans genes. They are incredibly small threads written with chemical letters and stored in the nucleus of all our cells. Cells are the tiny things that make up our bodies, like bricks make up a house. Cells find out how to behave from their genes, which make the cells work together to make a particular person and their characteristics—a boy or a girl, a short nose or a long one, dark skin or light skin, and many other features.

What our genes make can be seen in all living beings. How genes work and how our cells have managed to get their genes happen in a world too small to be visible, in the cells of our bodies. In order to investigate this further, let's take our MicroMachine. It will make us small enough to be able to watch cells and genes at work. We will see how they function, how plans are passed on from parents to children, and why we look a bit like and a bit different from our parents.

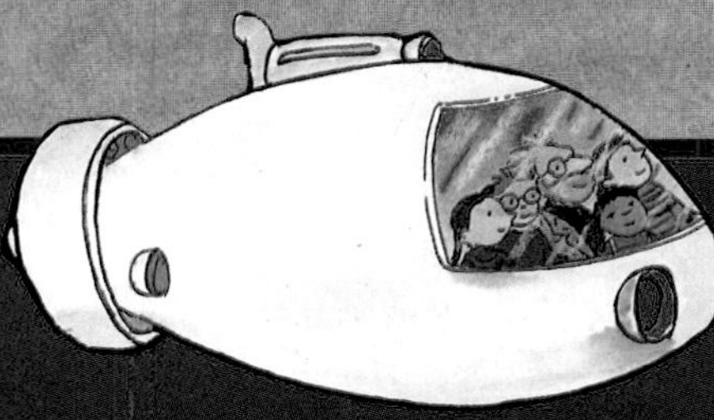

I am Professor Gene and the leader of this expedition. Just call me Gene. If you don't understand something—just ask me. One more thing! During our journey, we will see lots of strange things. Keep your dogs on their leashes—and let's go!

Cells make people

Imagine, a person consists of about ten trillion cells, which come in over 200 varieties! Most cells are so small that you could lay 100 of them side by side in less than one-tenth of an inch. Yet cells are really living beings. They eat food to get energy and create building matter. A new, young cell is born when a parent cell divides into two cells. When old cells die, new ones replace them.

Skin cells, for instance, fit together to form our skin; muscle cells fit together to form muscles; bone cells make our bones; and nerve cells and other cells supporting them make our brain and our nerves. We can sing, jump, think, and digest because we consist of different kinds of cells that do different jobs. Together they form an organism, keep it working, perform many functions, and protect it from germs that could cause illness. Every living being—every tree, every dog, every person—is the collective work of all its cells, which, most of the time, cooperate in perfect harmony.

What are cells made of, Gene?

Cells are made of many different substances, such as proteins, fats, sugars, and nucleic acids like DNA and RNA. Just as the joint efforts of many cells make a body, the interplay of these different substances makes a cell. In order to obtain substances to build cells, we need to eat food. Our stomach and intestines break down the food into smaller pieces, and our blood then carries the pieces to all the cells. Cells take out of the blood the specific things they need to build themselves and other new cells.

What is DNA?

It's incredibly thin strings found in each cell. On these strings, genes are lined up in certain patterns. Genes are recipes for proteins that the cell must build in order to act like a cell and stay alive. These recipes, or genes, are nearly identical in every person. That is why people are pretty much alike in most ways. We look like people instead of like dachshunds. However, some of our genes vary just a little bit from one person to another.

These minute differences are enough to keep us from all looking like identical twins. For example, some of us can jump higher because the muscles are stronger and the bones longer. Most clearly, however, the differences show in our faces.

There are also many invisible differences. Some children cannot drink cow milk or are color-blind. Some can remember a melody better than others, who are perhaps better at drawing pictures. Even our voices are different. If we disguise our voices, our friends may still recognize us on the telephone.

Do we also smell differently, Professor Gene?

Of course. However, human noses are not good enough to notice the different scents. A dog can do that much better. It sniffs at a shoe and knows whether it has been worn by someone it knows. To a dog, every person smells different—just as every person looks different to us.

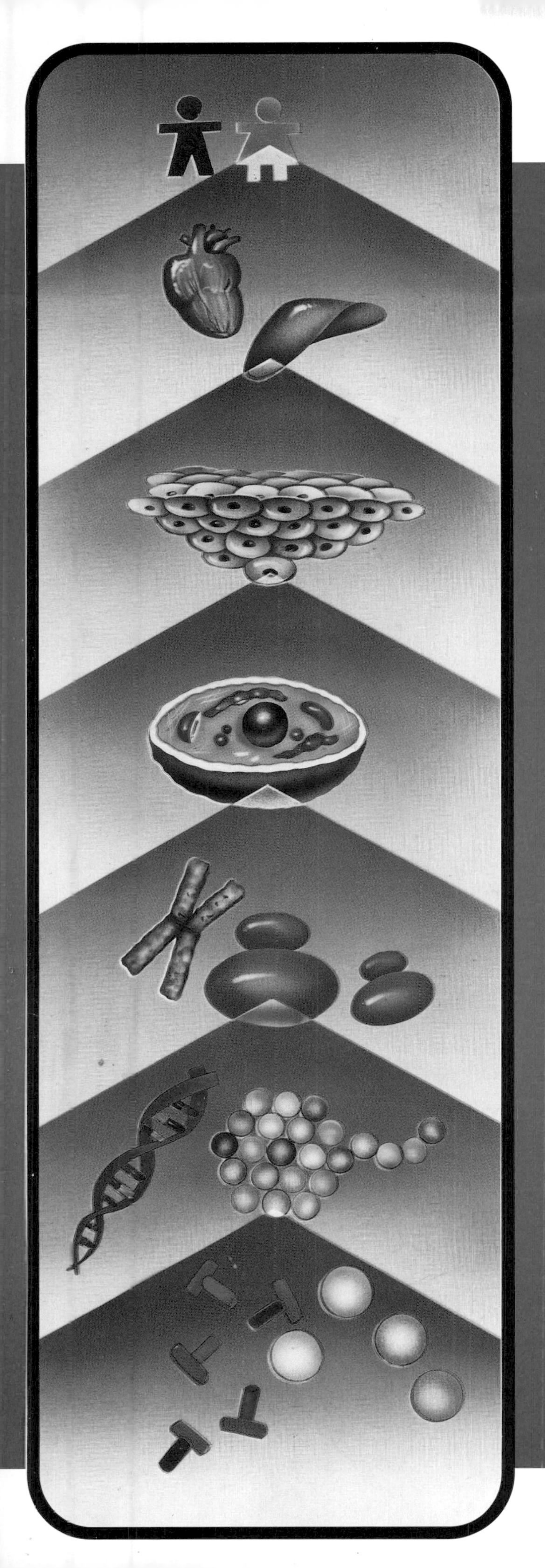

What we are made of?

People are made of their body parts.

Organs such as the heart, liver, or skin are made of cells forming a tissue.

Cell tissue is made of cells.

Cells are made of cell parts (organelles) such as chromosomes and ribosomes.

Chromosomes and ribosomes are made of large biomolecules such as DNA and proteins.

DNA and proteins are made of molecules such as amino acids and nucleotides.

Cells work!

Endoplasmic reticulum

Mitochondrion

Golgi apparatus

Nucleus

Cell skeleton

Ribosomes

Lysosome

In order to do its work, a cell needs genes and proteins. Genes give the directions for building proteins, which are the most important substances in a cell. Each kind of protein is made according to a particular gene. Cells need proteins as building blocks, glue, tools, pumps, copy machines, signal givers, slides, and packaging agents—for everything that makes a cell and keeps it working properly. The proteins made by a cell determine what functions the cell can perform, how it will look, and where in the body it will have to do its job.

Proteins also form the many different parts of a cell called organelles. Proteins and organelles make lots of fascinating things happen. For instance, the genes stored in the nucleus of the cell are constantly being copied so that new proteins can be made by the tiny protein factories called ribosomes. Old proteins are broken down by lysosomes. When a protein is supposed to leave the cell, it is made by ribosomes bound to the endoplasmic reticulum (ER). It then goes through a little bubble to the Golgi apparatus and can leave the cell. In the mitochondria, the food that has arrived in the bloodstream is burned. In this way, the cell gets its energy. You see, cells are bustling with activity.

Now all these skin cells, bone cells, muscle cells, or nerve cells can start working. They all know what to do. In a joint effort, they build a particular living being.

Do we grow bigger because our cells get bigger, Gene?

No, our cells always stay the same size. In order to make the body grow, the cells we consist of have to divide to make more and more cells. Bone cells make new bone cells; muscle cells, new muscle cells; and skin cells, new skin cells. Freshly made cells stick very closely together. They make, for instance, the cartilage, bones, and muscles in our nose, the skin that covers it, and the moist mucous membrane inside it.

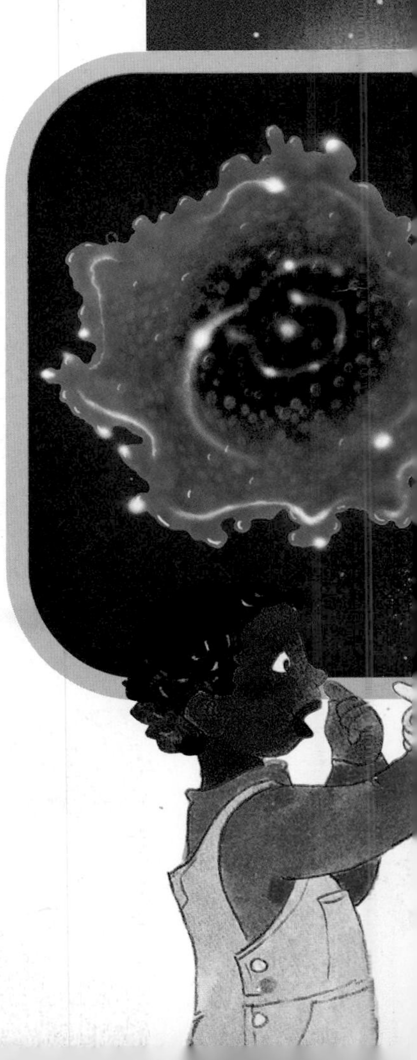

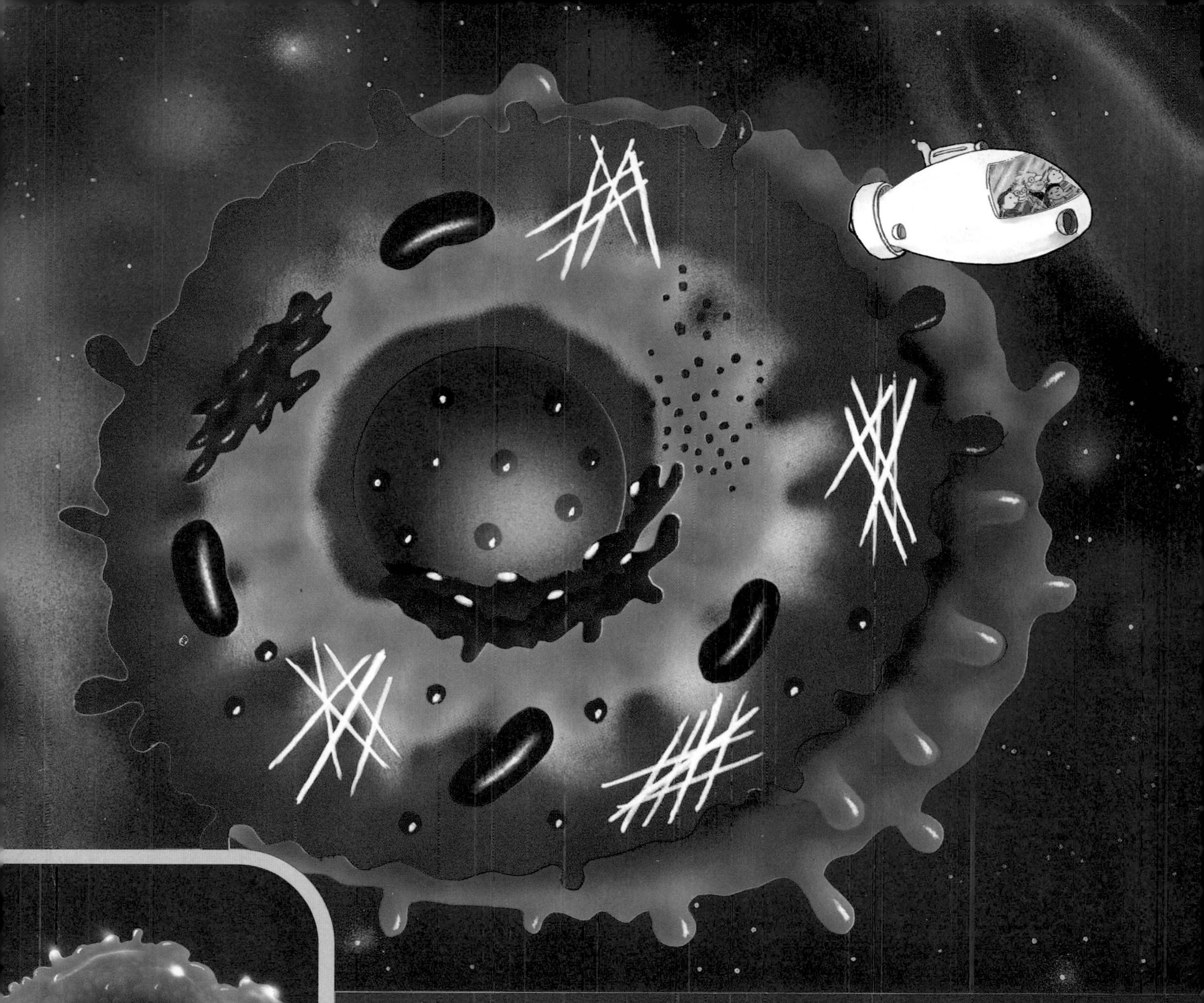

All our cells are connected with their neighbors in a variety of ways. Proteins are constantly moving about between cells and carrying messages from one cell to another. For instance, one protein will tell a cartilage cell that is helping to build the nose bone whether and when it should divide itself again. As long as a child is growing, the nose is growing, just like the other parts of the body, until it reaches the right size. In people with short noses, the cartilage has grown more slowly than in people with long noses. Their nose cells have divided less often and the nose remains smaller.

How does the nose know it should become a short nose?

The genes of the cartilage cells contain the plan. The genes even tell cells how often they should divide. There are very tiny differences in the genes of different people. These differences determine that a nose is long, short, or crooked when it has finished growing.

The *Y* makes the *guy*

On the inside, nearly all cells look alike. They are covered by a membrane. Inside, there are different little cell organs. We are especially interested in the nucleus of the cell. You see, it contains strangely shaped packages containing genes. These packages are called chromosomes. In the nucleus of (almost) every one of our cells are 46 chromosomes in 23 pairs. Here we see such a landscape of chromosomes.

Why do chromosomes come as pairs, Gene?

Each of us has inherited them from two persons, our mother and our father. One chromosome in each pair comes from our mother and one comes from our father. Each chromosome in the pair contains genes responsible for the same things. This is good. If a certain gene on one of the chromosomes of a pair is damaged, the healthy one on the other chromosome can take over.

Here is another drawing of the chromosomes. These smaller or larger X shapes appear only during cell division when the chromosomes are pulled apart.

Now you can see that the chromosome pairs vary in size. However, almost all of the chromosomes of girls and boys look the same. In girls, both chromosomes of a pair are of equal length and look identical. They all have an X shape. In boys, however, one of the 23 pairs contains chromosomes of differing lengths. The long one is called the X chromosome and the short one is called the Y chromosome. This tiny Y makes the guy.

How can the Y make a boy, Gene?

Chromosomes are nothing more than wrapped up genes. What matters is the genes themselves. The only difference between boys and girls is that boys have some additional genes on the Y chromosome. These are extra genes that girls do not have and do not need. They make sure that certain proteins are built in the cells of a boy. They will make a boy's body form all the organs that make him male.

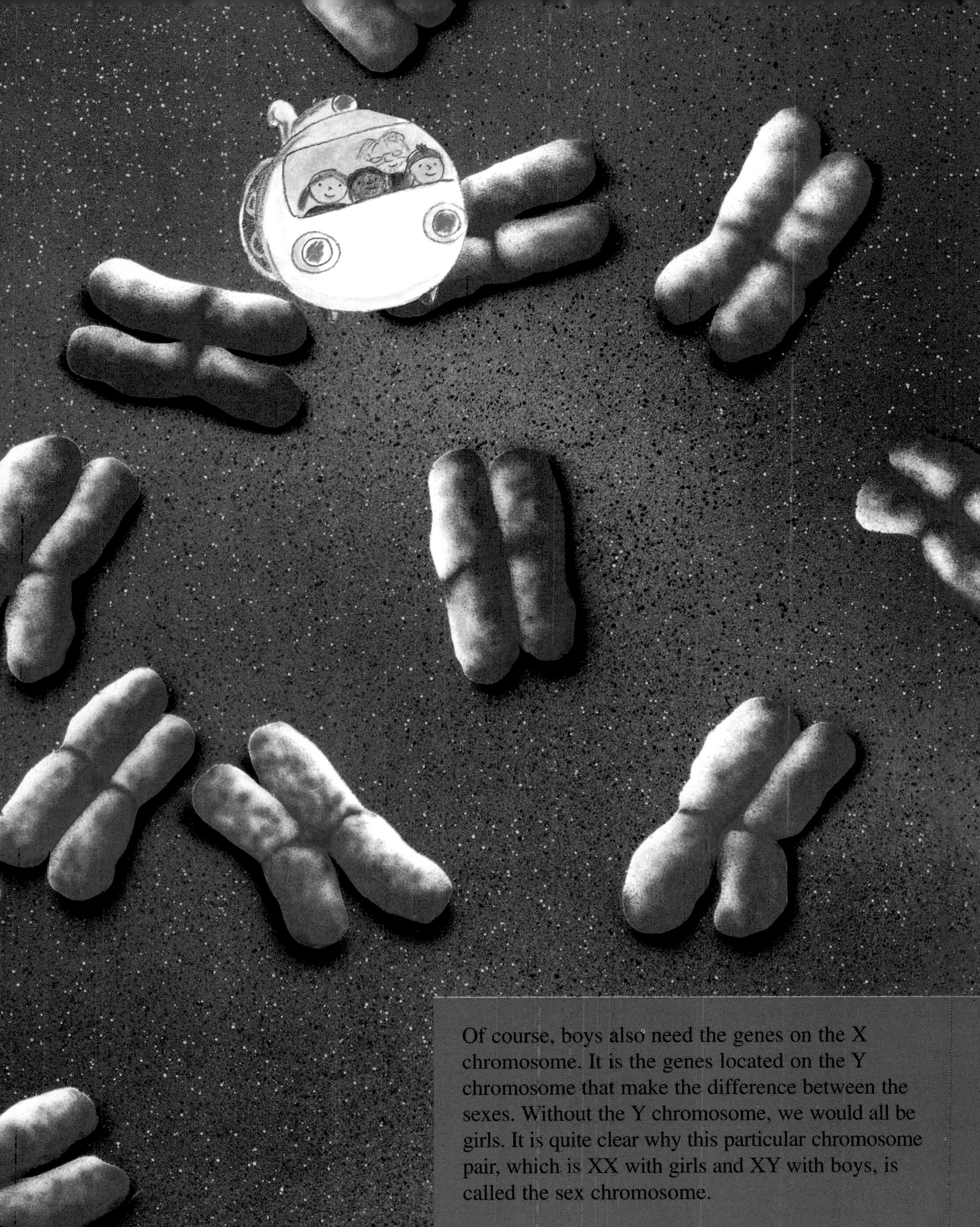

Of course, boys also need the genes on the X chromosome. It is the genes located on the Y chromosome that make the difference between the sexes. Without the Y chromosome, we would all be girls. It is quite clear why this particular chromosome pair, which is XX with girls and XY with boys, is called the sex chromosome.

Now it is time to take a closer look at genes. To do so, we have to make ourselves even smaller.

Genes and chromosomes

This is just one of the 46 chromosomes. When we make ourselves even smaller, we see it contains a jumble of threads. Now we can see that the threads are rolled up into tiny balls formed by special proteins called histones.

Finally, this is the actual DNA thread itself. It consists of two strands lying side by side and winding around each other like a spiral staircase. The rungs that connect them are made of two buddies. These are formed by four different chemicals called nucleotides that have been nicknamed A, T, G, and C. A gene is nothing but a little segment of the DNA thread containing several thousand nucleotides on one chromosome. The sequence in which the nucleotides are lined up on a gene tells the cell how to make a particular protein.

Can you see that the rungs are all made of nucleotide buddies—always GC, CG, AT, or TA? These are the only possible combinations. T would never connect with C or G, and C never with A or T. So if the staircase gets unzipped in the middle and if we had only half of the staircase, we could tell exactly how the other half needs to look.

If the staircase were split and each half given fresh nucleotides, all these As, Ts, Gs, and Cs would cling to their respective buddies and make a second half. As a result, two identical staircases would appear, both with the same nucleotide buddies in the same sequence. This is exactly what happens when a cell divides and duplicates its genes to give one copy to each of its two daughter cells.

There is another great thing about genes. Not only can they duplicate, but they also allow copies to be made that will be used outside the nucleus as patterns for making a particular protein. Let's take a look at how this works.

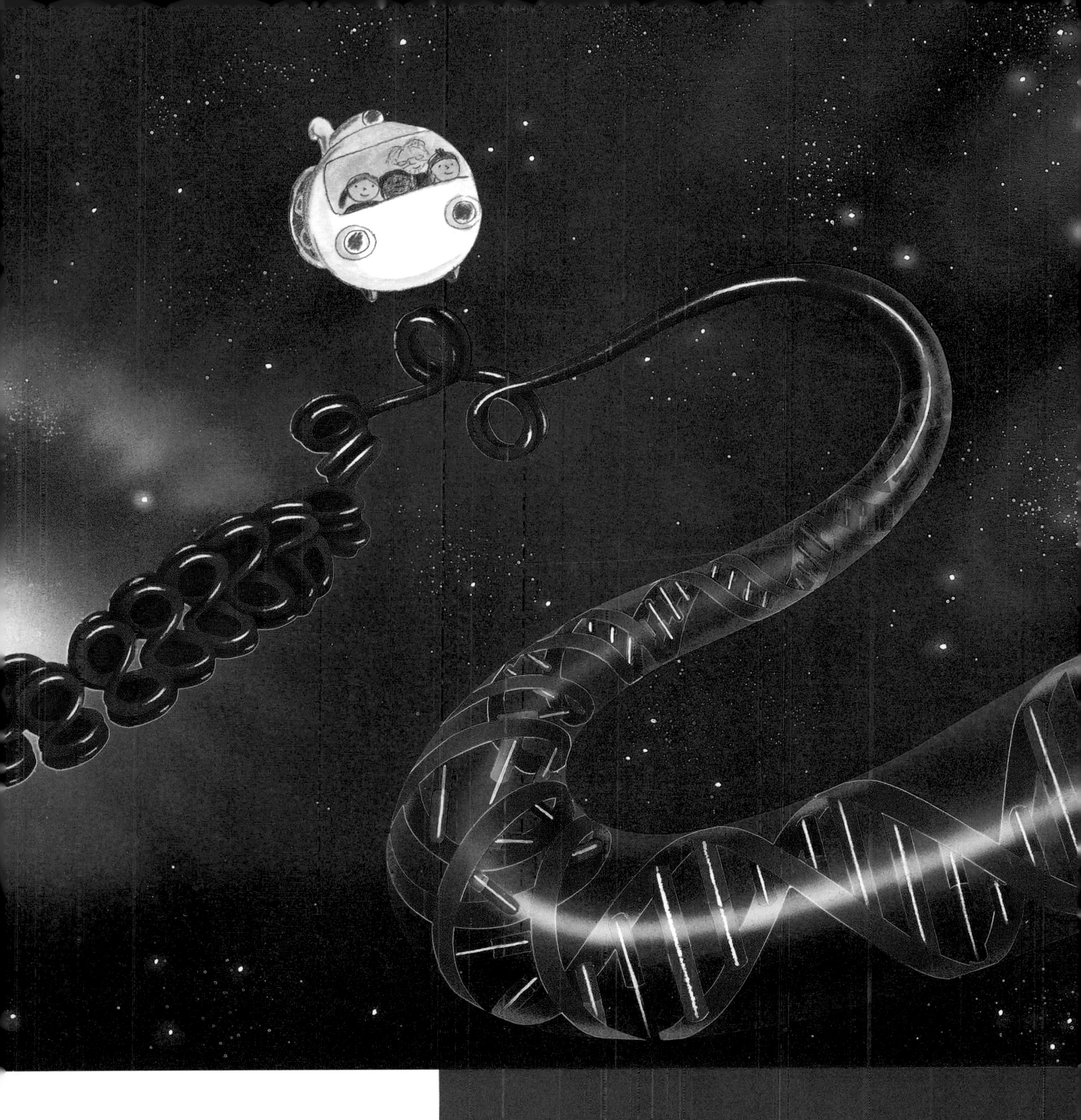

C A

G T

The nucleotides on the DNA ladder have a strict buddy system. C (shown in yellow) always and only joins with G (green), and A (red) with T (blue).

Proteins make people!

Exactly on the spot where a gene begins, the DNA staircase can be opened up a bit. The halves of the rung then become accessible.

Now special proteins (called polymerases) start doing their job. They always attach a T to an A, an A to a T, a C to a G, and a G to a C. This way a fresh thread of tightly connected nucleotides appears—something called messenger RNA. RNA does not contain the nucleotide nicknamed T. Instead, it contains one called U. U takes the place of T. The messenger contains a true copy of the gene—the plan for making a particular protein. The messenger RNA thread leaves the nucleus of the cell with its message. It goes to the protein factories in the cell, the ribosomes. Ribosomes look like double balls, and there are hundreds of thousands of them in every cell.

Now we can understand why different cells can make different proteins even though the same genes are present everywhere. It all depends whether or not a messenger RNA copy of a particular gene is made. To switch on a gene means making a copy of it. For instance, red blood cells would switch on the gene for hemoglobin by making a messenger RNA copy of it. Skin cells would switch off the hemoglobin gene by simply not making a copy of it.

The single-stranded messenger RNA is a true copy of the gene. However, the blue DNA nucleotide T is replaced by the violet RNA nucleotide U.

What are proteins made of, Gene?

Proteins are put together from 20 different building blocks called amino acids. A ribosome, one of those tiny protein factories in the cell, can read the gene on the messenger RNA. So the ribosome understands which and how many amino acids to line up in what order. The result is a unique protein thread that instantly starts to curl itself to form its particular shape. In the end, the finished protein leaves, ready to do its job. There are 100,000 different kinds of proteins working in the human body, each built according to a particular gene. Often, different proteins bind together into a larger particle. Ribosomes themselves are such particles that are made of many different proteins.

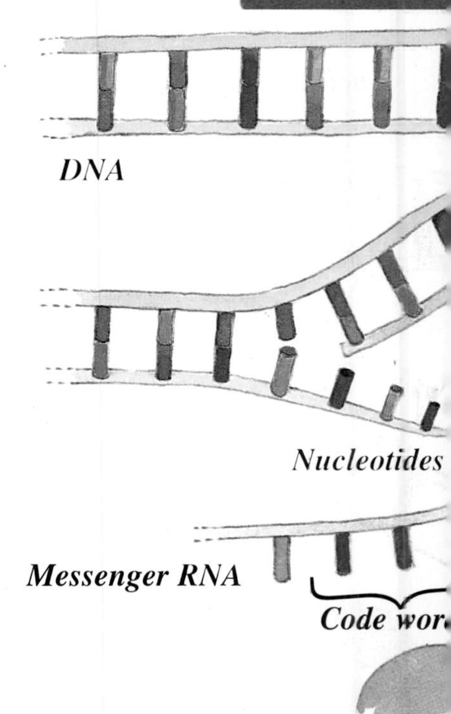

How can a human being develop this way?

Well, if we look only at a single process, it's quite hard to understand. Let's image that our genes and proteins are something like step-by-step instructions for making an origami figure from paper. At first, we have nothing more than a piece of paper and a list of instructions. We have no idea what the final result will be. We just fold the paper in the

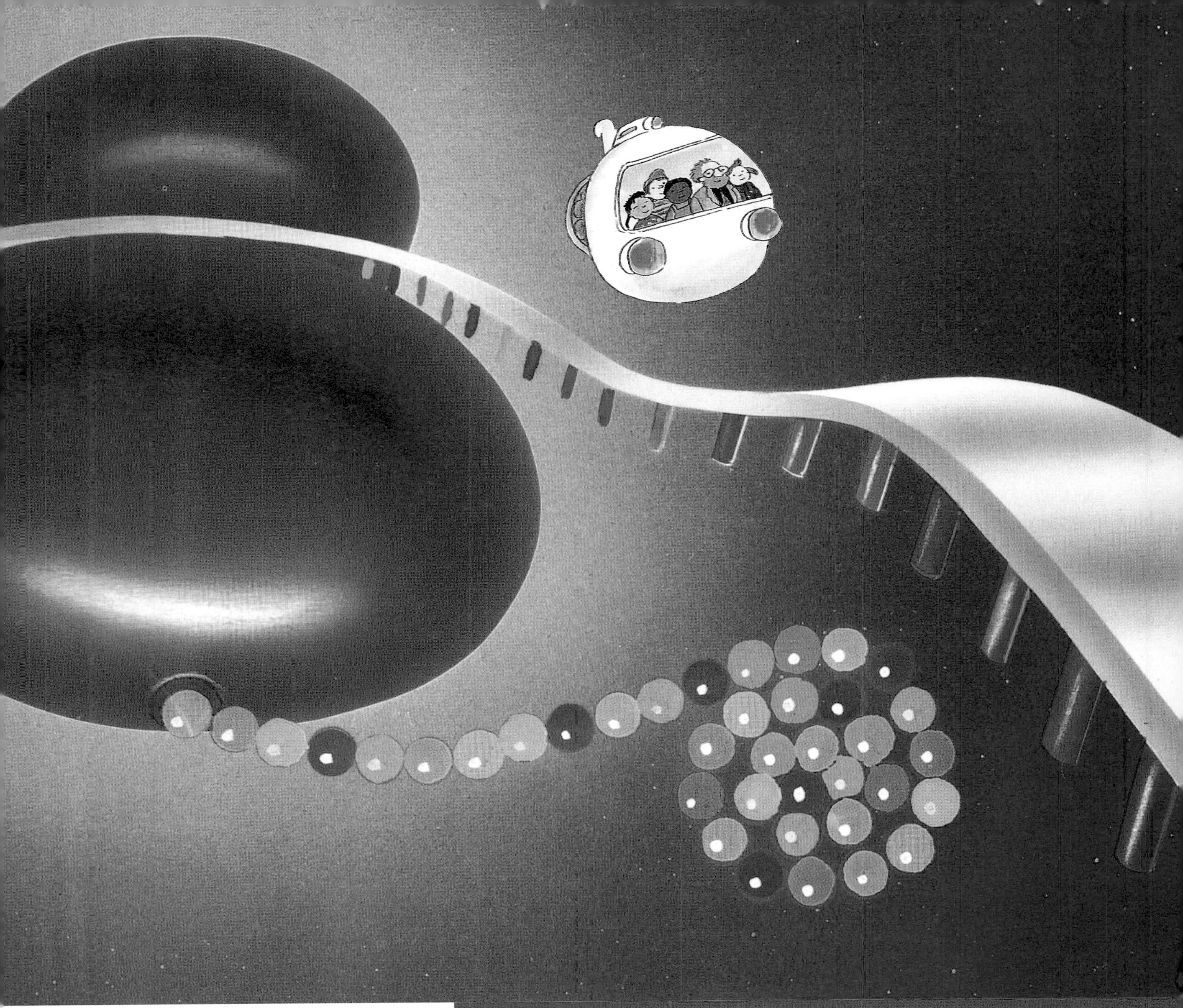

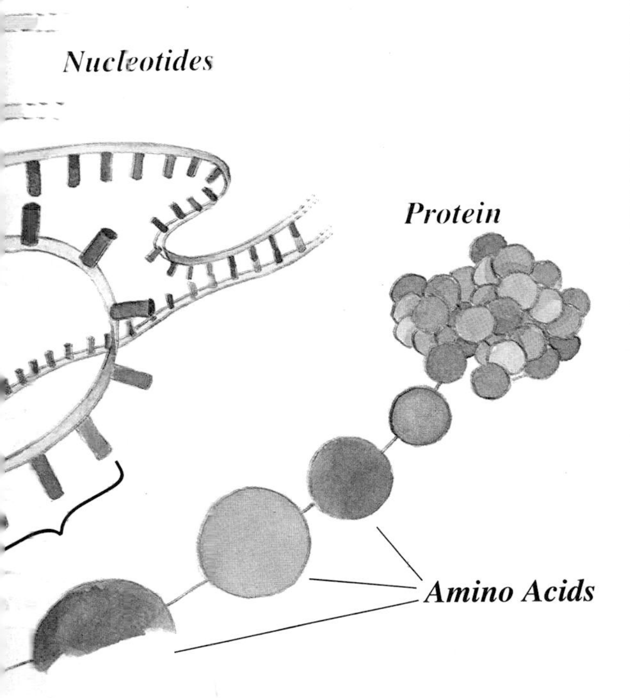

middle, then fold it along that line, then turn and fold there, et cetera. In each step, we just follow one simple instruction. Yet everything fits together so well that in the end we have, for example, a complete airplane. Look—it flies, even though we didn't even know what we were building at the beginning.

Similarly, we can imagine how the individual genes provide instructions first for proteins, then for cells, and finally for a complete, complicated living creature. The genes provide the plans, the proteins carry them out. This happens all the time in all of our cells. Everywhere, proteins of just the right kind, in just the right number, and at just the right time appear so they can interact in just the right way.

Cells **make proteins,** *proteins*

All of our cells are offspring of one single cell—the fertilized egg cell—that divided and divided again and made all the cells we consist of. Every time a cell divides, all the genes (together called the genome) are copied, and each daughter cell receives one copy. All of our cells contain the same genes. However, in different kinds of cells, different genes are switched on. Therefore, bone cells, for example, make different proteins than skin cells do and behave differently.

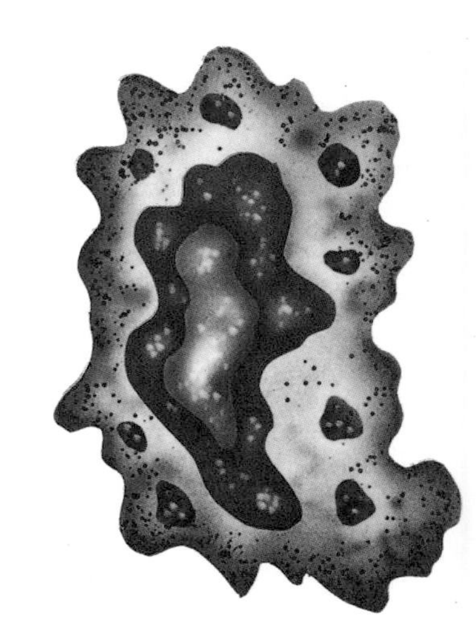

White blood cell

Proteins are so small that a million of them would fit into less than one-tenth of an inch. Every second our bodies are forming many billions of proteins. Altogether, human genes can make about 100,000 different kinds of proteins that each have different jobs.

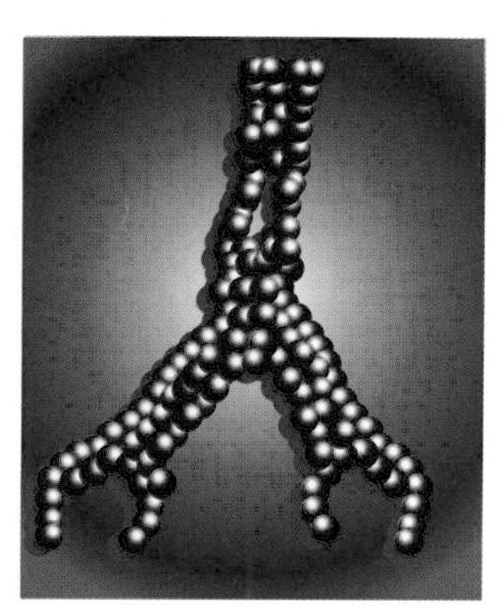

Proteins that fight germs are called antibodies.

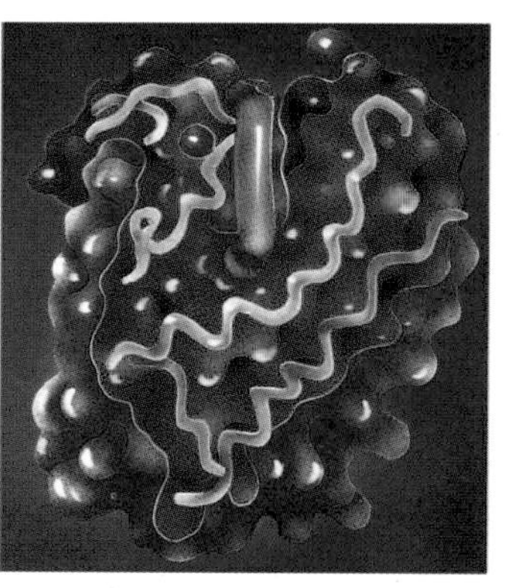

Hemoglobin is a protein that helps red blood cells transport oxygen.

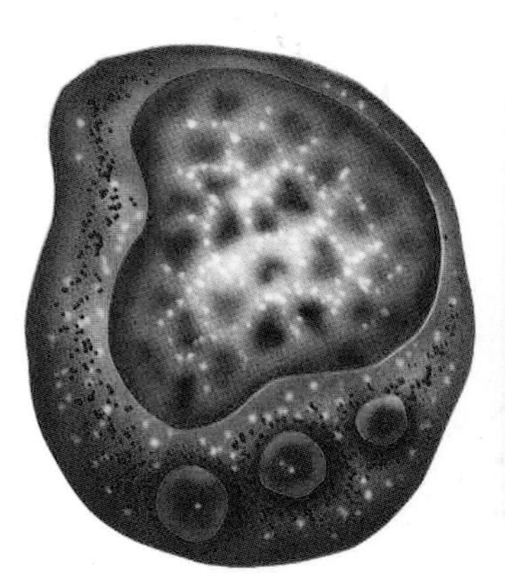

Memory cell

Some proteins are called housekeeping proteins. The genes that make them are switched on in all cells because all cells have to do some basic jobs. Housekeeping proteins make sure that cells can produce energy, divide, or deal with nutrients.

Other proteins form the skeleton inside a cell. Keratin, for example, is the main substance in our fingernails and our hair. Another protein, collagen, forms the elastic parts of the skin and cartilage and works together with minerals to build solid bones.

Some kinds of proteins make sure that particular cells can do their special jobs. Red blood cells, for instance, have to carry oxygen to cells everywhere in the body. So red blood cells contain a protein called hemoglobin that can capture oxygen.

Blood vessel cell

make cells

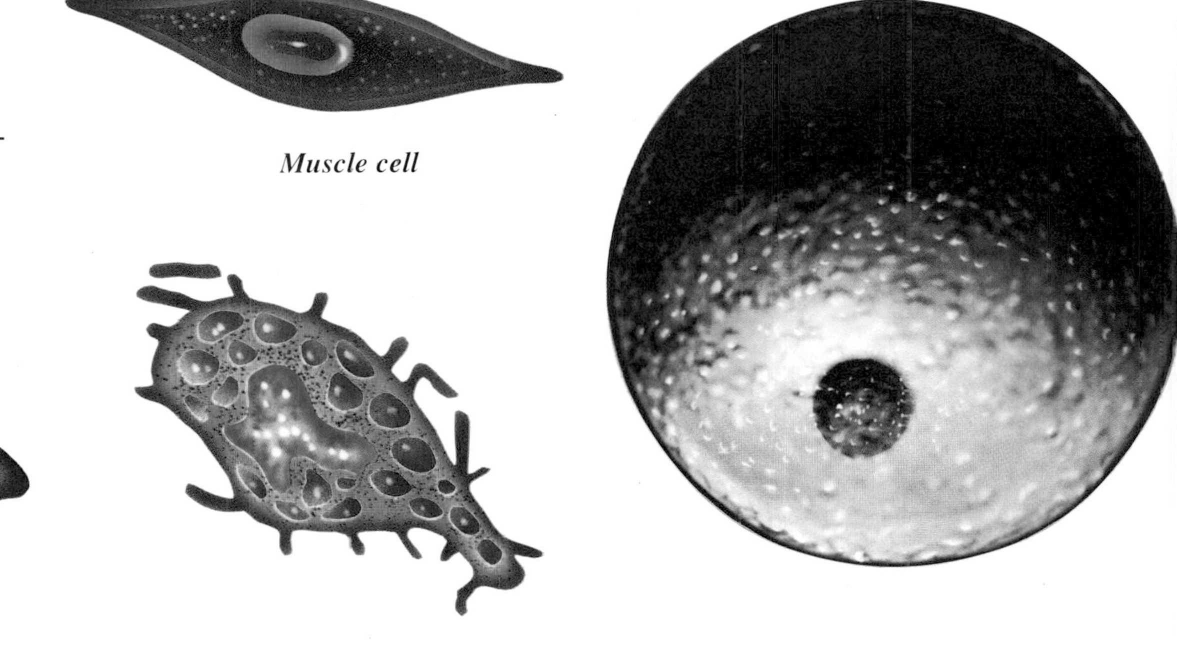

Muscle cell

Cell that makes hormones

Egg cell

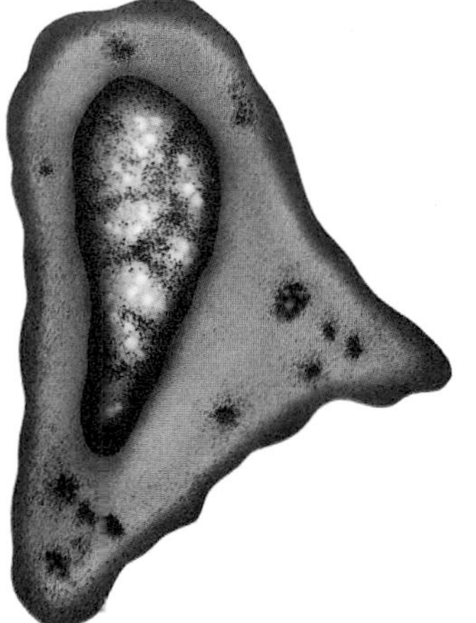

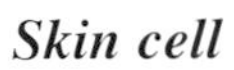

Skin cell

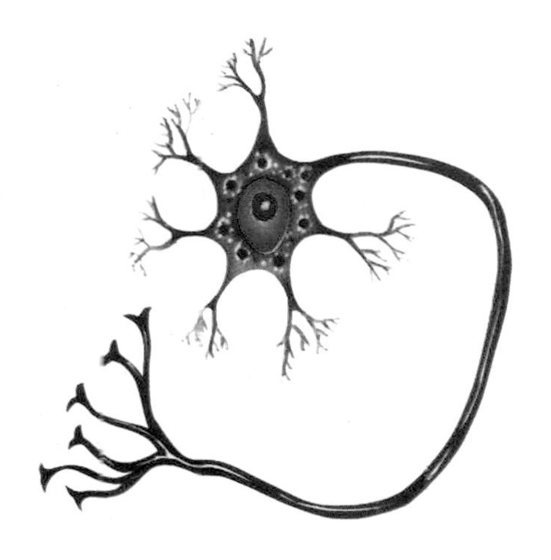

Nerve cell

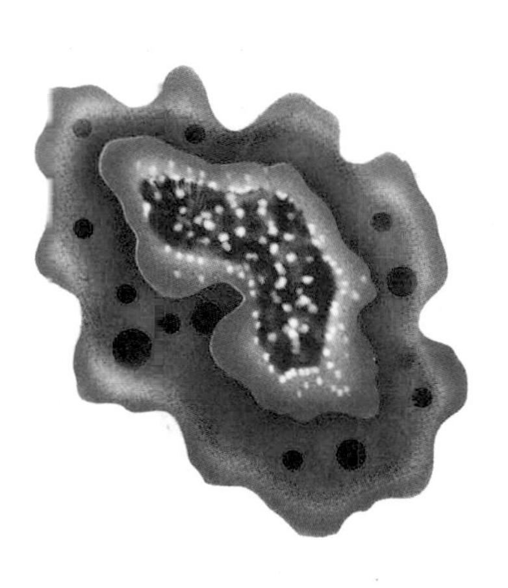

Bone cell

Other proteins, called enzymes, make sure that certain substances in the body are chemically combined. They also break down substances into smaller parts, for example when we digest food.

Still other proteins do the cell's work. Long, threadlike proteins called actin and myosin, for example, make our muscles contract. Tubulin is a protein that makes long tubes in every cell and can act as a track for little transport bubbles.

The proteins called hormones carry messages between cells. They tell a cell, for instance, to divide itself. The proteins called antibodies bring lethal messages. They stick to germs and make sure that bacteria and viruses will be destroyed.

Finally, many kinds of proteins guide and regulate all the operations within a cell. These proteins switch genes on or off. They make new proteins on DNA according to the building plans of the genes. They tell other proteins what to do and when to stop doing it.

Everything that happens in the body is connected in some way with proteins. For every kind of protein there is a gene—a pattern on the DNA thread telling every cell to make the very proteins it needs. Our 100,000 genes are all patterns for the 100,000 kinds of proteins. Genes and proteins all need to work together perfectly to create the living, growing human organism.

Different genes, *different* proteins

People all have nearly the same genes. Identical genes make identical proteins. Identical proteins make sure that all people have the same cells working in the same manner in more or less the same places in the body. So all of us have nostrils in the nose in the middle of our faces and not on the back of our heads, like dolphins have. We have eyes in the front and not on the side of our faces, like a rabbit has. We don't have a tail like a dog does.

Nose, ears, and lips—they all have a particular shape because the cells that build them divide and stop dividing at a certain time and a certain place. Thus, the cells form a cell tissue with a particular shape. In this way, our complicated internal organs are made.

Why don't people all look alike, Gene?

Every person has genes that are a little bit different. These genetic differences bring about the color of hair or eyes we have, the kinds of sicknesses we are prone to, whether we are a girl or boy, and whether or not we will get freckles. They determine the shape of the face and, of course, whether someone has a short nose or a long nose. In many cases, a lot of genes work together. Each of us has inherited his or her own mixture of genes, so our individual proteins are a little bit different. Thus, the cells work a little bit differently.

Groups of people have common characteristics. For example, children of Asian descent have skin cells above the eyes that grow more pronounced and divide themselves more often than that of children whose ancestors came from Europe or Africa. Thus, people of Asian descent have this special eyelid shape.

Skin color is another example. The skin cells of people with a darker color create larger amounts of a color substance or pigment called melanin. The color of hair and eyes is also determined by the genes, which we have at birth.

In the beginning was the *egg*

Now we know quite a lot about how genes function in our cells and how they make us what we are.

Where did we get our genes from?

All of our billions of cells, the muscle cells, the skin cells, and the nerve cells, have developed from one single cell—from the fertilized egg cell where each and every human or animal began. Unlike birds, the tiny human egg remains in the mother's womb until it has developed into a baby. A human egg cell is ten times bigger than a normal cell. A single egg cell develops only once a month in a woman's body.

An egg cell is special in other ways, too. At first, it contains only 23 single chromosomes. Since there are no pairs, the egg cell contains only one copy of each gene. It still lacks the 23 chromosomes containing the genes of the father.

The counterpart of the mother's egg cell is the father's sperm cell. The sperm cell, too, has only 23 chromosomes. It also contains only one copy of its genes. You see, separately, an egg cell and a sperm cell are quite helpless.

Until they come together?

Exactly. The tiny little things swimming there are the sperm cells. Unlike the single egg cell produced each month, there are many sperm cells. They are created in great quantities in the testicles of a man. Only one of them will be able to enter the egg cell and fuse with it. Once fertilization occurs, the egg cell will have 46 chromosomes—23 pairs. A new human being can begin to develop. One set of genes is from the mother, one set is from the father. The fertilized egg cell now has a double set of genes, just like all of the other cells in our bodies. It can now do what other cells can do—divide.

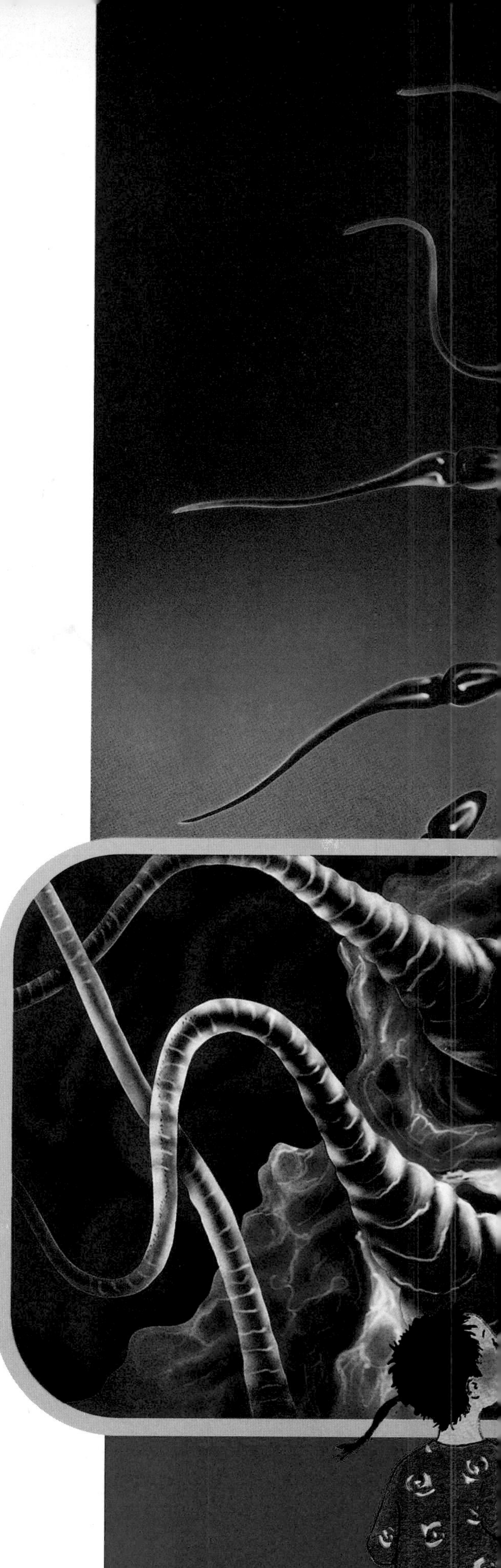

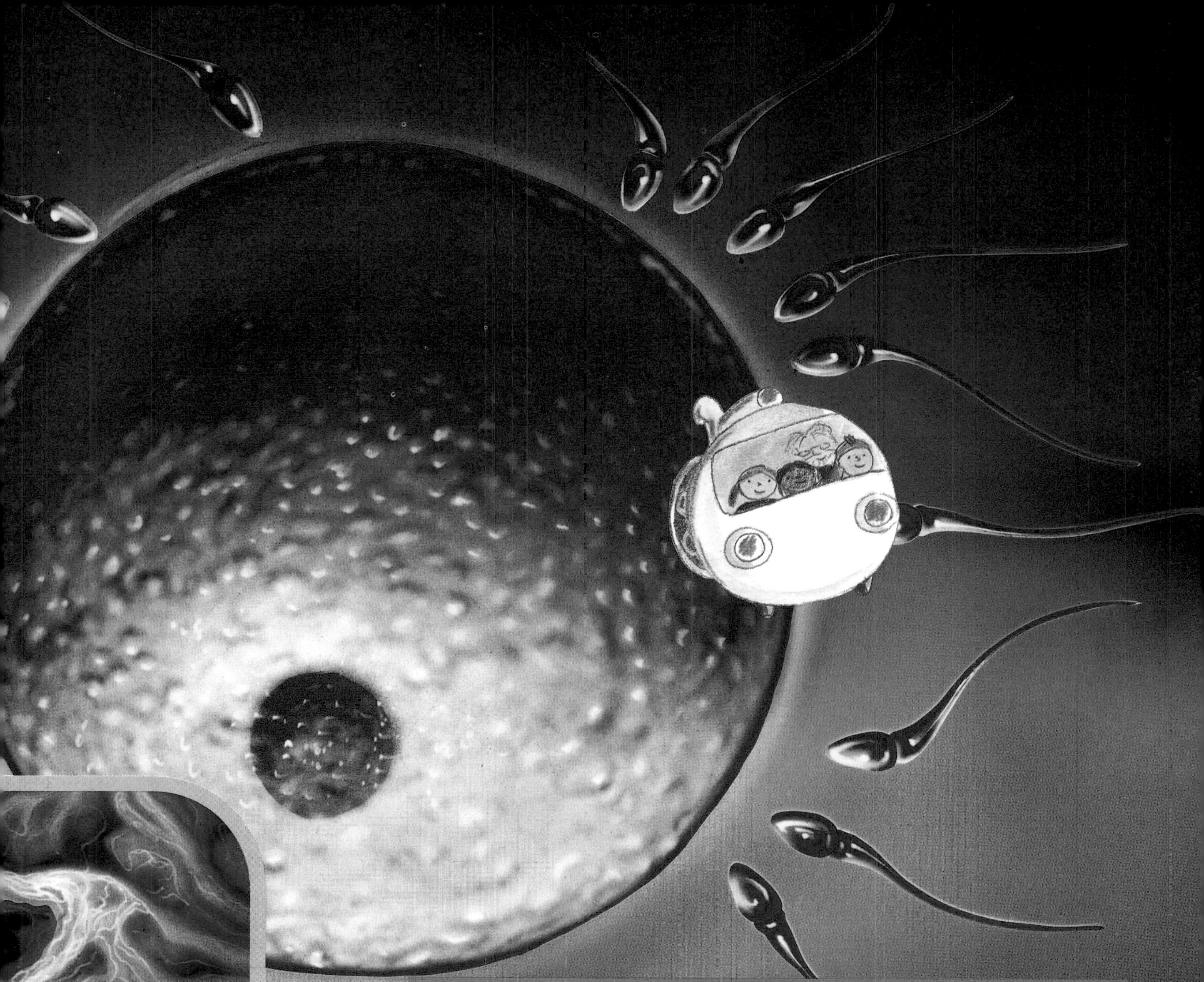

The daughter cells of the fertilized cell divide themselves again and again. Soon they will form the different kinds of cells that are needed to build a complicated organism. This fertilized egg is called an embryo. It contains in its genes all of the inherited characteristics the person will have—from hair color to nose shape.

Why doesn't the egg cell look like a tiny human?

It is still a cell. The same kinds of cells look pretty much the same in humans, dogs, and frogs. Our dog, for example, developed from a fertilized egg cell that looked just like the one I, a human being, developed from. Of course, a dog egg cell contains the genes of a dog. Dog genes make dogs, and human genes make humans. So the offspring of a living creature is always of the same species as the parents.

Boy or *girl* ?

A fertilized human egg cell contains more than just the patterns needed to build a general human child. The genes also determine specifics—whether it will be a girl or a boy, what color the eyes will be, approximately what size it will be, what kind of voice it will have one day, and also what kinds of illnesses it could develop in the future. However, these specifics won't develop for a long time.

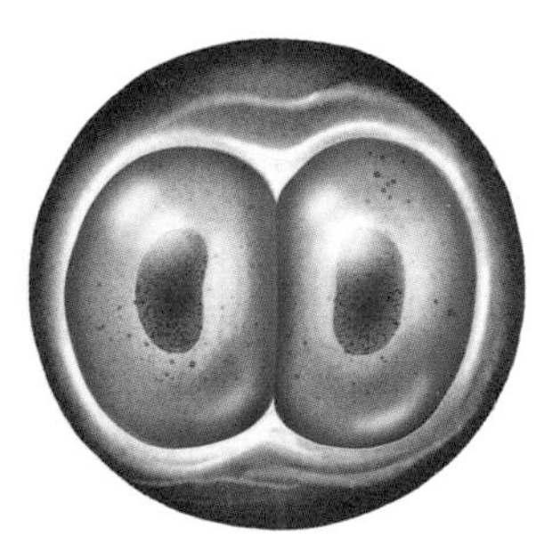

First, the egg cell begins to divide itself over and over again. In the beginning, the new cells are all identical. They form a solid ball and then a hollow sphere that slowly starts to push in. A few weeks later, cells that have different shapes and tasks start to appear. Some of them become nerve cells, others form the tissue that will later become the heart, while other cells are destined to become part of the intestines or the kidneys, for example. All of these cells have an exact copy of all the genes that first were in the fertilized egg.

How do boys and girls develop, Gene?

At first, all embryos develop in the same manner. Soon a very interesting gene is switched on—but only in the embryos that have a Y chromosome. This gene creates a protein that switches on a whole series of other genes. Now, particular cells are made that form the testicles. That is where the hormones are created that will turn the embryo into a boy and later into a man.

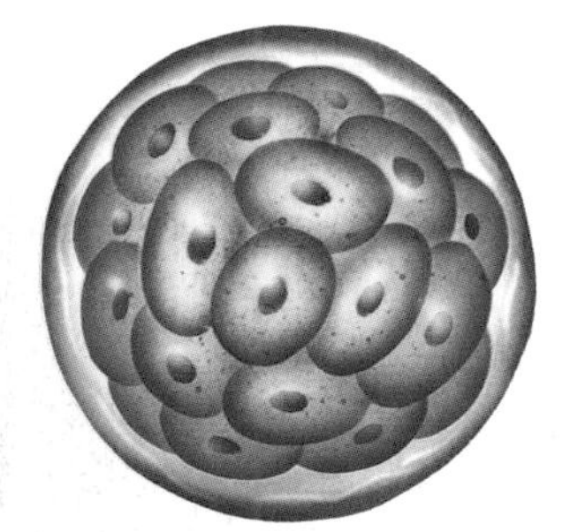

The gene that makes a boy is present only on the Y chromosome. When an egg cell is fertilized by a sperm cell that contains a Y chromosome, the embryo develops into a boy. If the sperm cell contains no Y but, instead, an X chromosome, the embryo will be a girl. The egg cell can contribute only an X chromosome, because a mother has only X chromosomes, no Y chromosomes. This way, the sperm cell of the father determines the sex of every child.

Nine months later, a baby is born, which will continue to grow for about 20 years. In the males, sperm cells are made; in the females, egg cells. In this way, people can pass on their genes to the next generation.

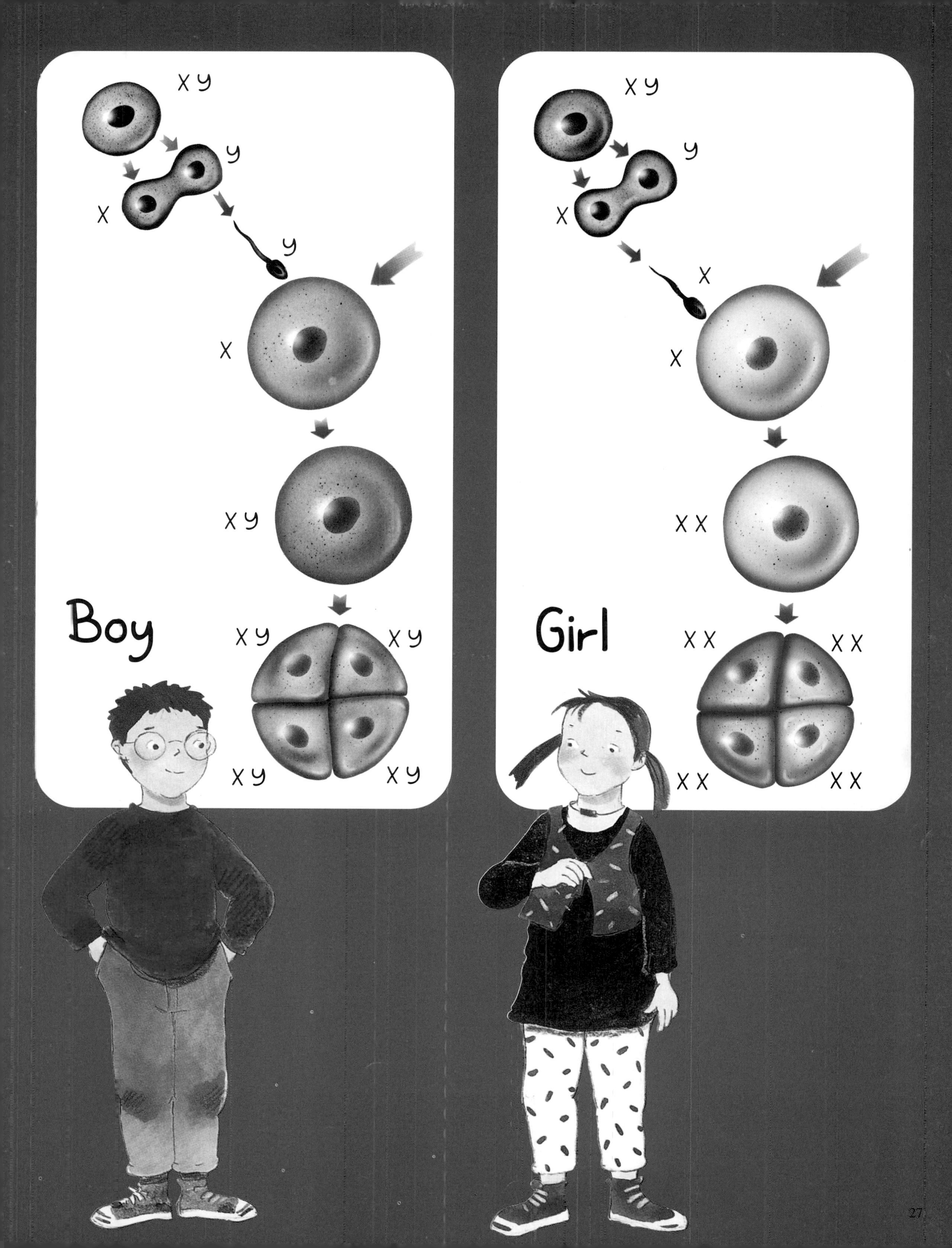

XY
Y
X
Y
X
XY
Boy
XY
XY
XY
XY
XY
Y
X
X
X
XX
Girl
XX
XX
XX
XX

Why do we need a *mother* and a *father?*

Living creatures that consist of only one single cell, such as bacteria, can multiply quite simply. The bacterium just divides itself. The two newly created bacteria each contain a copy of all the genes of the original cell. A bacterium does not have a mother or father, only one parent cell. Our body cells divide and multiply themselves just like bacteria do. If we need new skin cells or new liver cells, they simply divide until enough new skin cells or liver cells have been made.

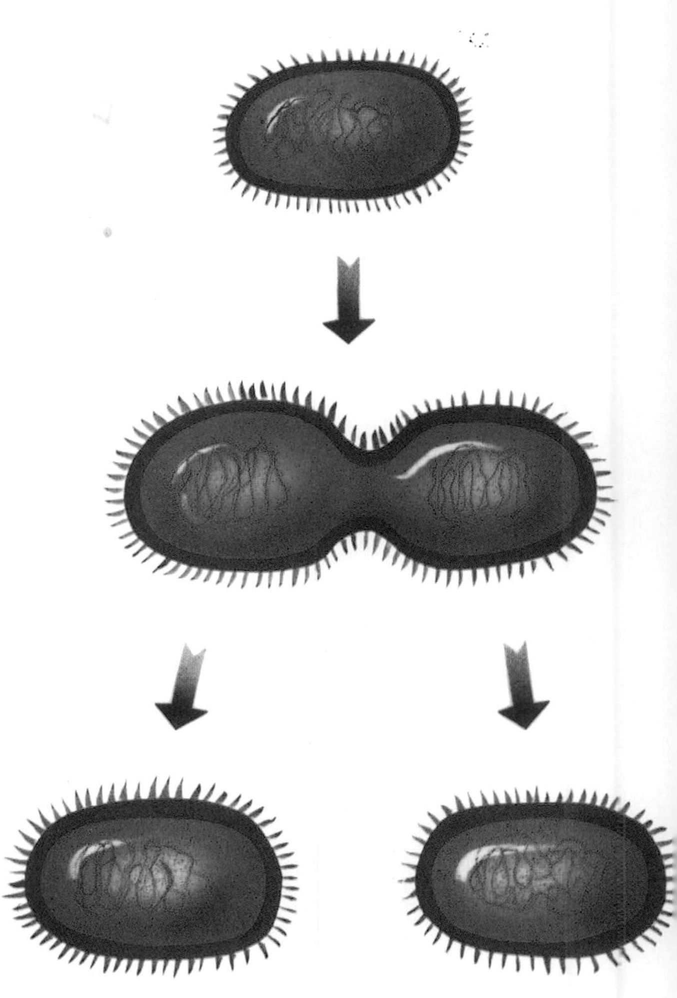

For humans (and for most animals and plants), things are not as simple as they are for single-celled creatures. First, we are made of many billions of cells. Secondly, our cells have become specialists. Although a skin cell contains all genes to make a human, it has switched off all the genes a skin cell wouldn't need. A skin cell can divide to make only fresh skin cells, and nothing else.

We cannot simply divide ourselves in the middle and—voilà!—become two children. People, trees, and dogs need to have cells that are not specialized These nonspecialized (undifferentiated) cells can lead to the creation of all possible kinds of specialized cells—skin cells, nerve cells, or liver cells, for example.

The fertilized cell in the mother's body is this undifferentiated cell.

Can there be a baby without a father, Gene?

No. An unfertilized egg cell cannot divide alone. It needs the genes from the sperm cell. Otherwise, children would be born that have exactly the same genes as the mother, the grandmother, and the great-grandmother. People would be identical females but of different ages. Nature has arranged for children and parents to have different genes.

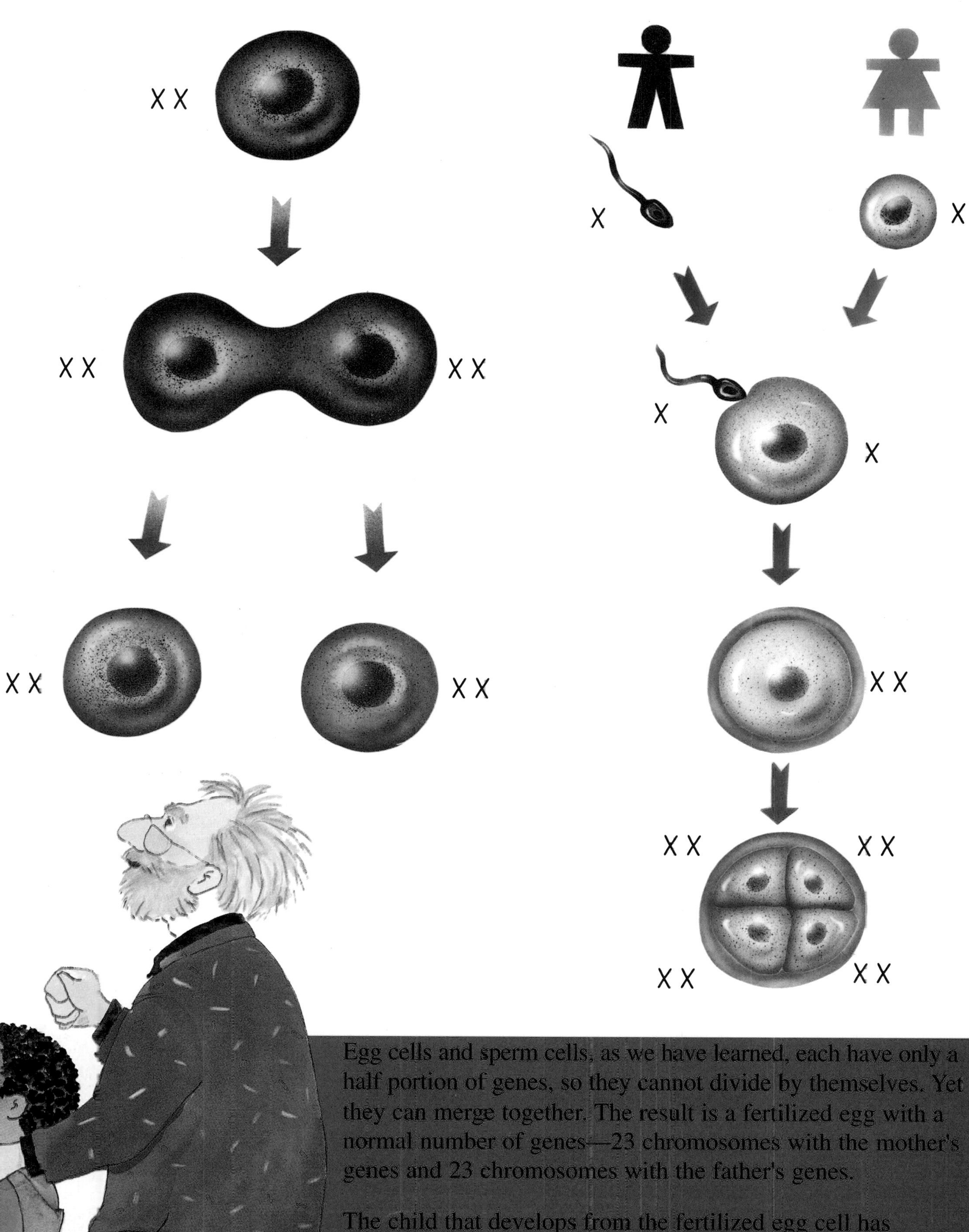

Egg cells and sperm cells, as we have learned, each have only a half portion of genes, so they cannot divide by themselves. Yet they can merge together. The result is a fertilized egg with a normal number of genes—23 chromosomes with the mother's genes and 23 chromosomes with the father's genes.

The child that develops from the fertilized egg cell has received genes from two people, from the parents. Each of the parents has inherited their genes from two other people, the grandparents, and so on.

Genetic *mixture*

The two very special kinds of cells, egg and sperm cells, are called germ cells. The way they come together to make a new child is called sexual reproduction.

Do germ cells develop from our other normal cells?

Yes. However, they develop in a very special way. Germ cells develop in the ovaries of the mother and the testicles of the father. Let us take a look at how sperm cells are made. At first, the cell in the testicles that makes sperm cells has, quite normally, a double portion of genes, 46 chromosomes in 23 pairs. Its offspring, the sperm cells, are supposed to have only a single portion of genes—23 individual chromosomes. How does that happen?

Why not divide the chromosomes?

Well, this would be simple enough. Put one set of chromosomes into one resulting sperm cell, the other set into another one—that's it. No, it's not that easy for the cell that makes sperm cells.

Before the final cell division, the chromosomes are mixed thoroughly. The two chromosomes of a pair move quite close together so that the same gene segments are opposite each other. Then the chromosomes are broken apart at the same places. The resulting ends unite themselves with the ends of the opposite chromosome. This occurs many times over again in almost all chromosomes. Only after this crazy mixing has occurred will the cell that makes sperm cells divide. Now the chromosomes can be divided among the fresh sperm cells. In this way, each child inherits from the father a random mixture of genes that the father had inherited from his parents.

The same chromosomal mixing occurs in cells that make egg cells. Hence, each fertilized egg cell—and the person that will develop from it—has his or her own unique set of genes. The person has a mixture of genes that has never before appeared on this planet and will never show up again. We have our own particular characteristics and features that make us different from our parents and our brothers and sisters.

This is how germ cells are made. All this happens in the nucleus of a sperm- or egg-producing cell.

1. This is, for example, a sperm-producing cell. Just like any other body cell, it contains 23 pairs of chromosomes. To see things more clearly, let's only look at one pair.

2. The chromosomes of one pair embrace each other. Both contain genes that are in charge of the same tasks and located at the same position on the chromosome. Now the DNA strands in the chromosomes break at the same position and the loose ends cross over to join with the loose DNA ends of the other chromosome.

3. After having swopped genes, the chromosomes drift apart again. Now each chromosome has received its unique and random mixture of genes.

1

2

3

4

4. Now the cell starts making sperm cells. When dividing, one of the freshly mixed-up chromosomes of one pair goes to one future sperm cell, the other one goes to the other sperm cell. So the genes within a particular sperm cell not only differ from father's genes; they also differ from the genes of all other sperm cells ever to be made!

The same thing applies to egg cells. Germ cells don't have chromosome pairs, but single ones. Only when they meet will the resulting fertilized egg cell have chromosome pairs—and grow to make a baby.

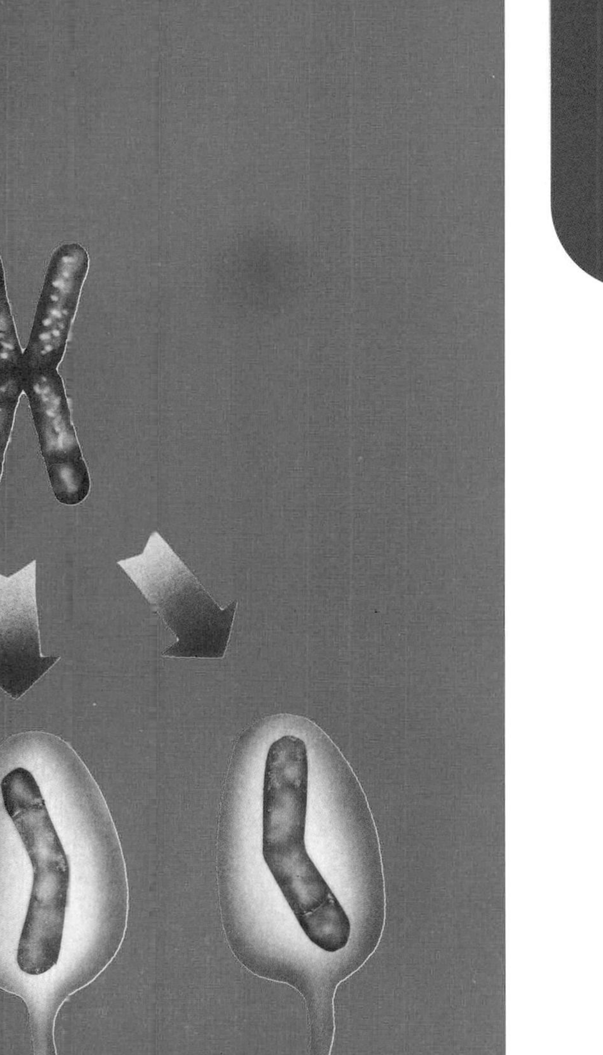

Why is there an almost equal number of boys and girls, Gene?

Let's figure it out together. The cell that makes egg cells has two X chromosomes. When it divides, both resulting egg cells must have an X chromosome. Remember, a woman makes only one egg cell each month. The egg can receive either one of the two X chromosomes. So the resulting egg could be called

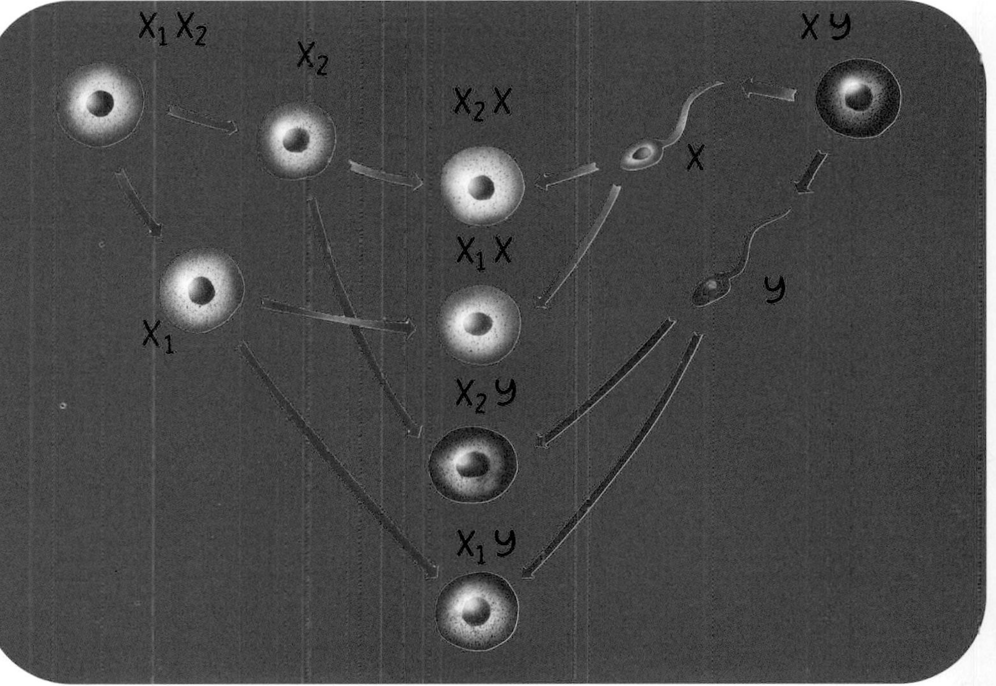

Egg-X1 or Egg-X2, depending on which X chromosome it contains.

The cell that makes sperm cells has an X and a Y chromosome. When it divides, half of the new cells receive an X and the other half a Y chromosome. We will call them Sperm-X and Sperm-Y.

When we place the egg cells and sperm cells together, the following combinations are possible:
Egg-X1 and Sperm-X result in XX, a girl;
Egg-X1 and Sperm-Y result in XY, a boy;
Egg-X2 and Sperm-X result in XX, a girl;
Egg-X2 and Sperm-Y result in XY, a boy.

You see, half of the combinations will give boys, the other half girls. Whether parents will have a boy or a girl is therefore a matter of chance. Some families have only boys, other families have only girls. Altogether, almost the same number of boys and girls are born in the whole world.

A double portion, *Please!*

All of our cells—except, of course, our germ cells—have a double portion of genes. On both chromosomes of a pair, the genes responsible for a particular protein are located at the same spot. As a rule, these genes are both switched on or off at the same time.

Usually, these genes are identical. In some cases, genes come in two different versions called alleles. For example, genes are responsible for the color of hair. One allele, perhaps the one that came from the mother, says, "Make the hair black!" The allele that came from the father says, "Make the hair blond!" Both genes then go to work. What color hair will the child have?

A mixture of blond and black hair?

That is impossible. All of the cells in the root of a particular hair would have to decide to use a particular allele. The cells in another root would all have to choose the other allele. This choosing doesn't happen. Instead, the child will probably have black hair. This is because genes for black hair are normally stronger than genes for blond hair. Therefore, we call the genes for black hair dominant genes. They produce different amounts of protein in each and every hair root. Genes that are present but do not dominate, like the blond hair genes, are called recessive.

Sometimes, one of the two genes is faulty and cannot create proper proteins. If important proteins are missing in the body, we can become ill. Usually, the healthy allele is able to take over the job and keep us healthy. A faulty gene is also called recessive—as long as it doesn't cause an illness.

Are all illnesses the result of sick genes, professor?

No, certainly not. We get sick from colds or other infectious illnesses because of viruses and bacteria, not defective genes. However, genes can play a role in the way our body deals with an infection. Illnesses that are based on faulty genes are called hereditary diseases—they have been inherited from the parents.

Twins and other siblings

Now we understand why a child can have a short nose like grandfather or blonde hair like grandmother even though the parents have long noses and black hair. If two genes for blond hair or a short nose come together in the child's DNA, those genes, although they had no visible effects on the parents, can determine how the child looks.

Many thousands of genes are present in different versions. Each time an egg cell is inseminated, a completely new mixture is created. Hair color is only one simple example. Just think of all the different human characteristics—nose shape, body size, voice, or even such a complicated thing as musical talent. We all differ from each other in many thousands of ways. For many characteristics, not one but dozens or even hundreds of genes are involved. It is all the result of a unique mixture of genes we inherit from our parents, grandparents, great-grandparents, and all our other ancestors.

What about twins, professor?

Of course, twins also have different genes than their parents. Only in the case of so-called identical twins are there two people who have exactly the same genes. This is because identical twins come from the same fertilized egg. They develop as twins when the first tiny ball of undifferentiated cells divides into two separate halves. This happens very rarely. Only then do twins have the same genes and look so much alike that it is hard to tell them apart. They are always the same sex and have the same hair and eye colors. Quite often, they even have the same habits—even if one of the twins is adopted by another family and grows up separated from the other twin.

A living being that has exactly the same genes as another one is called a clone. Identical twins are clones that have developed in nature by chance.

Of course, some twins have different genes. Such twins develop when not one but two egg cells ripen in one month. If more than one egg cell becomes fertilized at the same time, very normal brothers and sisters with different genes will develop. They may have the same birthday, but they will look different and may even be of a different sex. These twins are called fraternal twins.

X

X

X X

X X

X X

X X

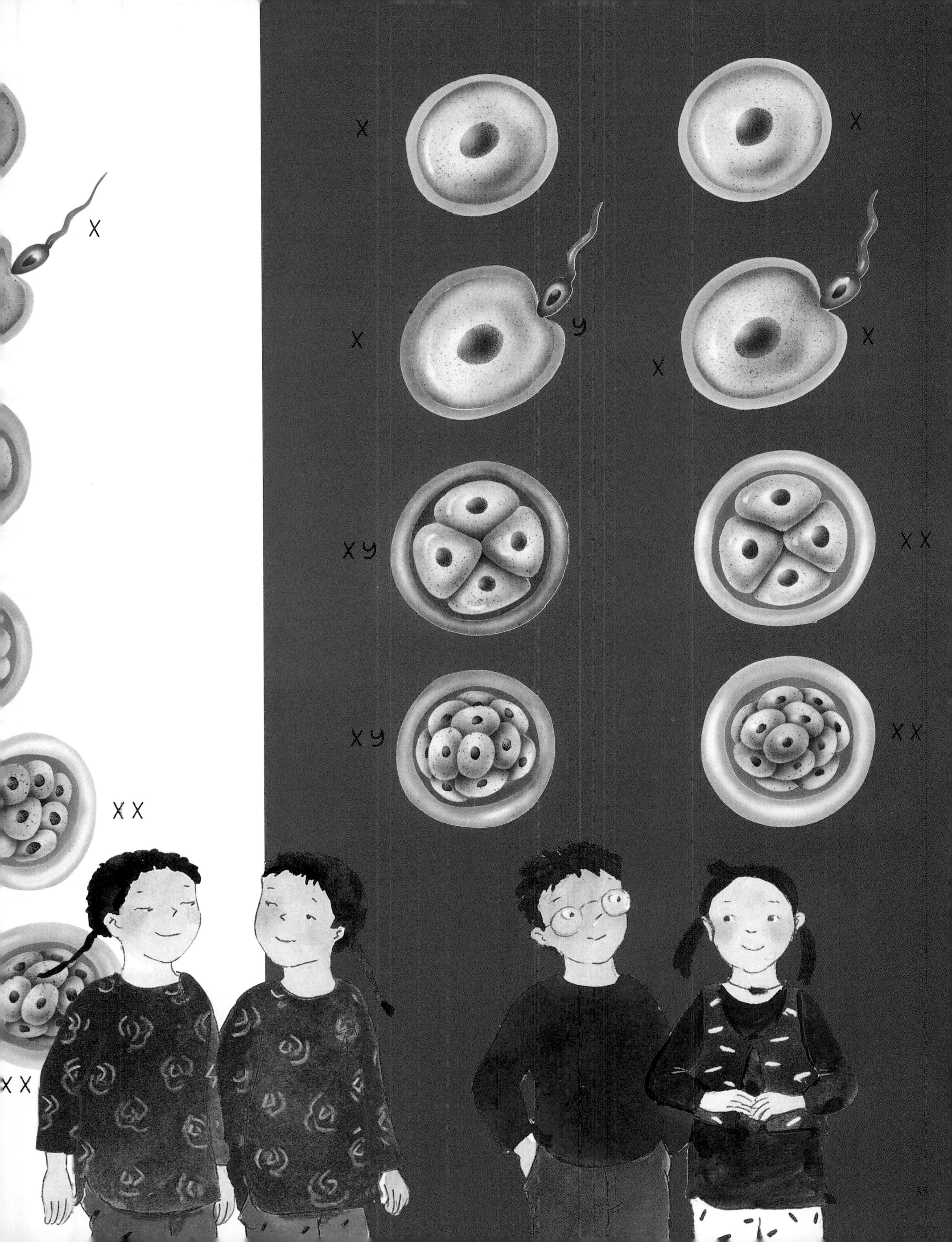
X
X
X
X
y
X
X
X y
X X
X y
X X
X X
X X

Sick genes

Living beings like humans and animals are fortunate to have received a double set of genes. For one thing, this makes us all a bit different from each other. Secondly, for almost every gene we have a reserve gene in case one should be damaged.

Does one faulty gene make any difference?

Well, in most cases, faulty genes are recessive. This is because genes are simply there to produce proteins that do something. If one of the genes that is supposed to make a particular protein is faulty, it would either not make proteins or make useless ones. The other, healthy gene normally does work properly. All the necessary proteins can still be made and will do the whole job. This is like an airplane in which one pilot suddenly gets sick. As long as the other pilot can do the job, the plane will arrive safely at its destination. The passengers probably would never learn that one of the pilots was sick—he or she was, in a sense, recessive. Fortunately, it is seldom the case that both pilots fall ill or that both genes we inherit are faulty.

In some diseases, one faulty gene does make a difference. For example, what if the protein that should be made is responsible for clearing the body of poisons? If only half the necessary amount of protein can be produced, more and more poison will accumulate—just like garbage would pile up if half of a city's garbage trucks were out of order. In such cases, patients will suffer from a milder form of a hereditary disease even if they still have one healthy gene.

Are there genes that have no reserve gene?

Unfortunately, yes. Do you remember that boys, but not girls, have an X chromosome and a Y chromosome? Well, on this X some very important genes are located. They make proteins that cause blood to clot in an injury. Without these proteins, blood just will not stop flowing even from a small wound. This disease is called hemophilia. Can you imagine why most hemophiliacs, or bleeders, are boys?

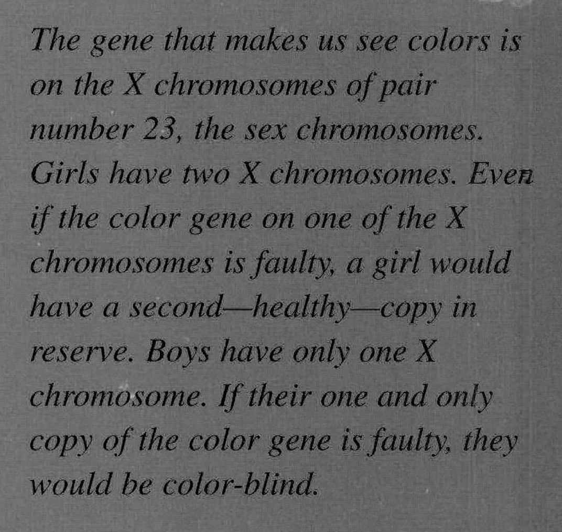

The gene that makes us see colors is on the X chromosomes of pair number 23, the sex chromosomes. Girls have two X chromosomes. Even if the color gene on one of the X chromosomes is faulty, a girl would have a second—healthy—copy in reserve. Boys have only one X chromosome. If their one and only copy of the color gene is faulty, they would be color-blind.

X X

Girl

X y

Boy

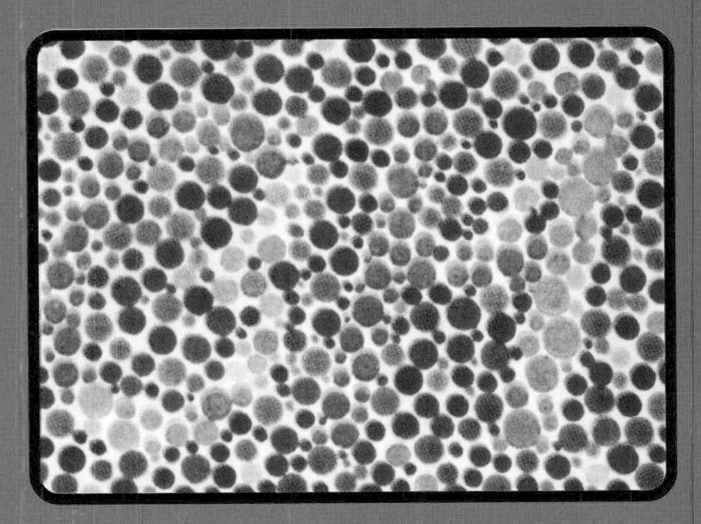

Look very closely at this picture. Can you recognize the "CH" patterns? If you read "31" instead, you might have a problem in telling colors and you might want to ask your parents whether you have color-blind relatives.

Do girls have a healthy blood-clotting gene in reserve?

Exactly! Girls and women have two X chromosomes that contain exactly the same genes, including a blood-clotting gene. Boys, however, have a Y chromosome instead of a second X and therefore have no reserve genes. If a boy's blood-clotting gene is faulty, he will suffer from hemophilia.

By the way, some other important genes are located on the X chromosome. For example, the gene that lets our eyes tell the difference between red and green is on the X. If that gene is damaged, the result can be color blindness. Boys are more often color-blind than girls simply because boys have only one copy of that gene, while girls still have a copy in reserve.

Mutants among us!

Cells are highly skilled in making copies of DNA. Every second of our lives, an incredibly large number of cells are dividing, making copies of billions of genetic letters. Only rarely does an error occur that makes a gene in a particular cell unreadable. Usually, this does not cause troubles. First, a reserve gene exists. Secondly, millions of other cells of the same kind can take over the sick cell's duties.

Genetic errors mostly matter when they happen in the germ cells. Egg cells and sperm cells can pass on all their genes, including the faulty one, to all the cells of a child—even to its germ cells. Such a genetic error is called a mutation. It is a change in the genes and will be passed along to future generations.

In many African Americans the red blood cells are not shaped like rubber boats but like sickles. These sickle cells are more fragile than normal round cells but can protect people from malaria, an infectious disease. Imagine, all this is due to a single wrong amino acid in the protein called hemoglobin.

What happens to mutated genes?

Usually, mutated genes are unreadable and thus faulty. Again, that doesn't really matter as long as a second, healthy gene acts like a copilot. Still, a mutation can be passed along to children, just like a healthy gene.

Sometimes mother and father each have a healthy and a faulty version of the same gene. Their child has a good chance of receiving at least one healthy gene. In the worst case, the baby receives two faulty genes—two sick pilots. Even though both parents are perfectly healthy, the child will be ill or become ill later. Hundreds of hereditary diseases exist. Most of them are hardly curable, and patients will require life-long medical treatment.

Are mutations always harmful, Gene?

Not at all. Mutations can also be very interesting and helpful. Actually, we are all mutants since lots of harmless genetic changes have occurred in our ancestors, resulting in things like short noses, red hair, or specially shaped earlobes. Such mutated genes create more variety in the types of people around us. We all are mutants—we look differently, which I think is fine. After all, we want to recognize people at first glance and not at first smell.

Some mutations can also create other useful changes. For example, it is helpful for people who live in hot, sunny places to have darker skin. Dark skin can withstand the sun's rays better than light skin. In dark skin, the cells make more of a protein that creates melanin, which

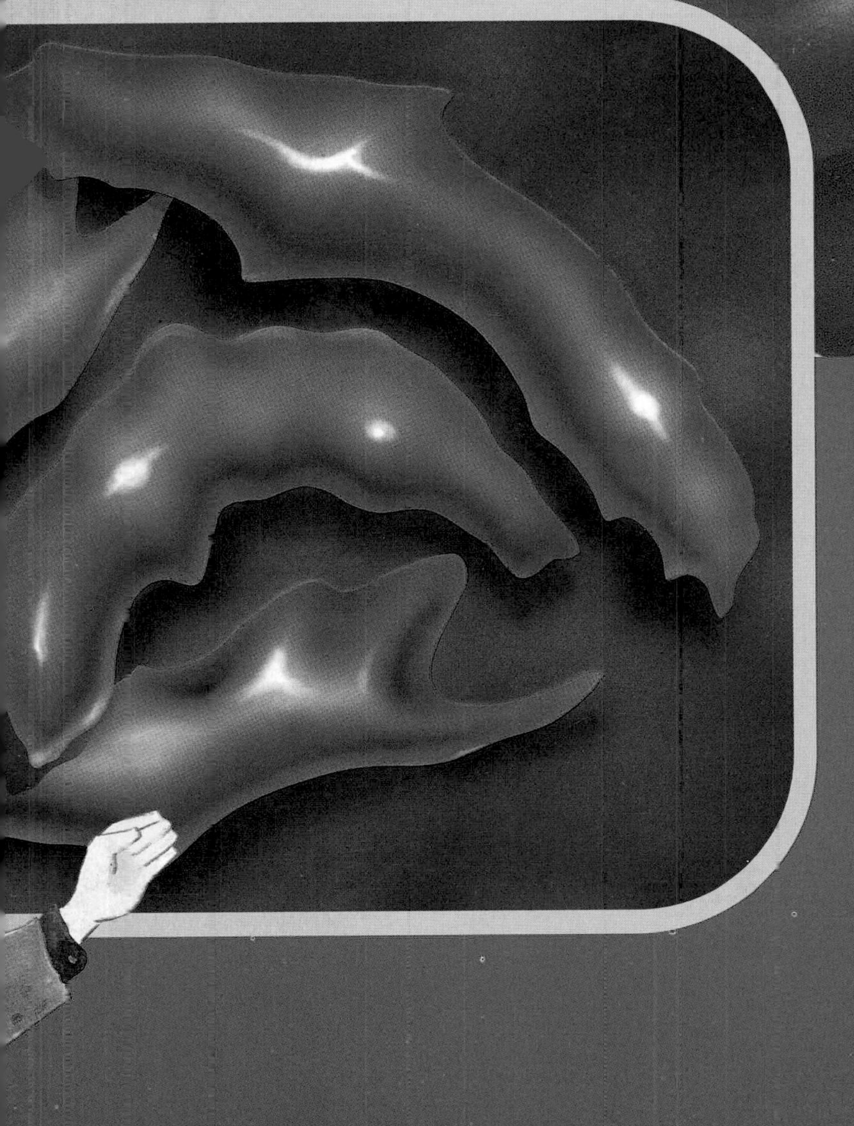

protects us from sunburn. On the other hand, for people in countries with little sunshine, light skin is more useful. It can take in more sunshine, which is needed to create certain vitamins.

People—and all other living beings—with favorable mutations pass on their genes to their offspring. It would take many generations until such useful changes would spread. Mutations can allow living beings to become adapted to different environmental conditions.

Genes are not everything

We are at the end of our journey through the world of genes. Now we understand the saying that someone has the eyes of his father, the nose of her grandmother, or color blindness from his grandfather. In reality, we don't inherit eyes, noses, or anything else. We inherit the genes that tell our cells which eye color to make, which way to shape a nose, and which proteins to produce that let us tell red from green.

Genes, however, are only responsible for the characteristics of a person that can be inherited. A human being is far more than just that. Take identical twins for example. They have identical genes and look quite alike. Nonetheless, each twin has his or her own special personality—just like everyone else on this earth. Each of us feels, thinks, and acts a little differently. This we owe to our brains.

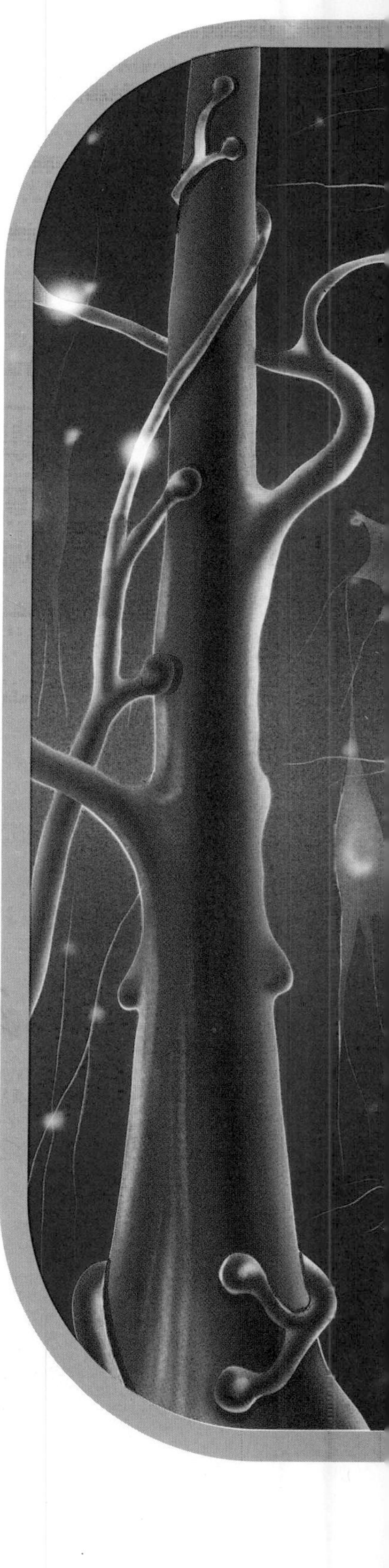

Is there a gene for intelligence, Professor?

No. The brain is unimaginably complicated. For characteristics such as intelligence or musical talent, many hundreds of genes play a role. We will probably never find out which genes work together in which ways to make one child better at arithmetic than another, another child learn a melody more easily, or another get along with other people better.

Most important of all, the brain is not something determined solely by our genes. Genes roughly dictate the number of cells in our brains. They do not, however, tell the nerve cells within the brain how many other nerve cells to connect with. That's what makes a difference. If we use our brains intensively, more connections between the individual nerve cells are created. That is what makes us more clever. Each of us has his or her own certain talents and weaknesses. Of course, only a few people will become geniuses like Mozart or Einstein. However, we all can make the best of ourselves and our talents.

That is something very special that only human beings can do. We can think about ourselves. We can make plans for our future. We can even investigate how a single little cell can become a complete person such as you and me, and how genes make all this possible. Our genes provide the plans to build our bodies, including our brains. It is up to us, however, what to make of ourselves.

What do you remember?

Every living creature consists of

a) at least one cell
b) at least two cells
c) at least 46 cells

a) is correct. Living creatures that consist of only one cell are called unicellular organisms. Examples of this are bacteria, some algae, and even baker's yeast. Other organisms such as trees, ants, and human beings have innumerable cells of many different kinds. Such cells could never exist on their own but must work together in an organism.

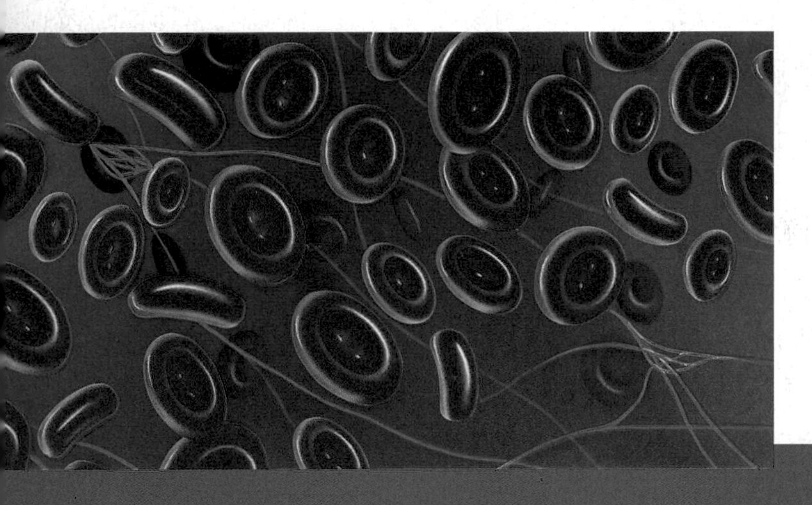

People all look different because we

a) have different genes
b) have different parents
c) eat different things

a), b), and c) are correct. Each person has his or her own particular mixture of genes inherited from the parents. Of course, what we eat also plays a role. People who eat too little or too much will have very different figures.

People have almost all our genes

a) in two versions
b) in as many versions as there are cells
c) in 46 versions

a) is correct. For most genes, both versions (which come from mother and father) are identical. Translated into proteins, they make the same proteins. Some genes have different versions that have different functions.

The two versions of a gene are called

a) a double helix
b) chromosomes
c) alleles

c) is correct. Chromosomes look like short noodles or dumplings in the nucleus of a cell. DNA ladders are packed in them. A double helix is the shape that the DNA ladder takes when its two threads wrap around each other.

The cells that form the cartilage of the nose and the cells in the roots of hair have

a) different genes
b) the same genes with only particular genes switched on

b) is correct. The genes in the different cells of our body are all the same. These ten trillion cells all come from the same fertilized egg cell. The nose cartilage cells, though, use different genes than those used by hair root cells. Different kinds of cells have to do different things. They switch on different genes and thus can produce different proteins.

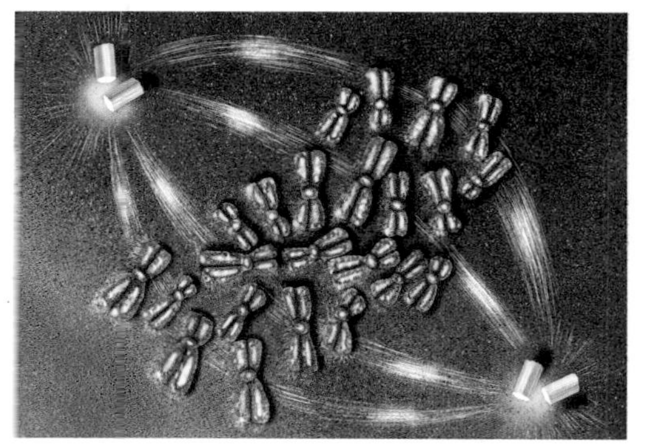

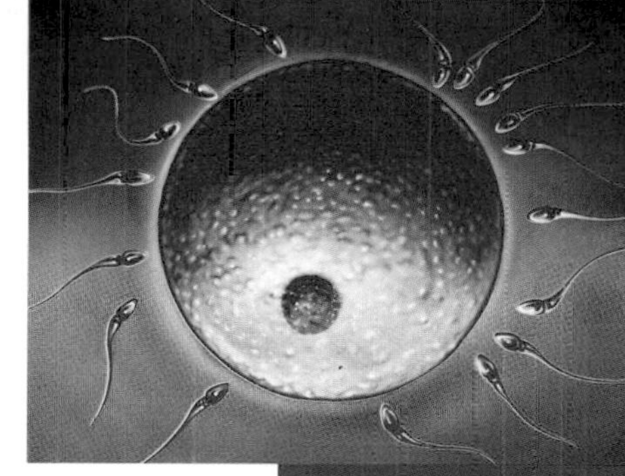

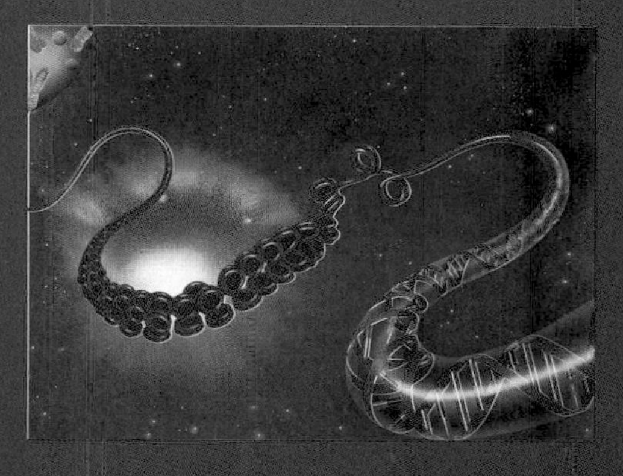

A particular gene is switched on

a) when a copy is made of it
b) when it is doubled
c) when it drifts out of the nucleus

a) is correct. The copy—the messenger RNA with the copy of the gene—travels to the ribosomes, the protein-producing factories of the cell. That is where all proteins are made. A gene for which no copy is made has no effect. It remains switched off. During cell division, genes will be doubled but not switched on. Genes always stay within the nucleus. Only the copy of the gene, the messenger RNA, leaves the nucleus.

Amino acids come in

a) 20 different kinds
b) four different kinds
c) about 100,000 different kinds

a) is correct. Using 20 different kinds of amino acids, our cells can build the 100,000 different kinds of proteins we need. Four different kinds of nucleotides are the building blocks of the DNA.

Egg cells and sperm cells are called

a) germ cells
b) embryos
c) sex chromosomes

a) is correct. Egg cells and sperm cells carry the germ of life within them. When they merge together, the egg cell becomes fertilized. An embryo is a fertilized egg cell that divides and grows, producing a living being. The sex chromosomes contain genes that cause boys and girls to be different.

A fertilized human egg cell that contains one X chromosome and one Y chromosome will result in

a) twins
b) a boy
c) a girl

b) is correct. The genes on the Y chromosome make the embryo build the male sex organs.

Glossary

Alleles different versions of a gene. For example, the genes for blond hair can have alleles for black hair.

Brain the central control system of the body. It consists of billions of nerve cells. It receives and sends messages along a network of nerve fibers to nearly all parts of the body.

Cell tissue made of cells sticking together.

Cells tiny living beings. Hundreds of different kinds of cells form the human body and help to keep it alive. The main task of a cell is to make proteins. Cells eat food to get energy and create building matter.

Chromosomes packages of DNA containing the genes. In the nucleus of almost every one of our cells are 46 chromosomes in 23 pairs.

Clones living beings that have exactly the same genes as other beings. Identical twins are natural clones.

DNA short for deoxyribonucleic acid. Genes are lined up on these incredibly thin strings, shaped like twisted ladders, that are found in each cell.

Dominant genes versions of genes that are used to make proteins.

Egg cells contain only 23 single chromosomes. In order to divide and form all the different cells an organism needs, a female egg cell needs to be fertilized by a male sperm cell.

Enzymes proteins that help combine and break down chemical substances.

Fertilization takes place when a male sperm cell fuses with a female egg cell.

Genes sections on the DNA. Stored in the cell nucleus, they are the recipes for making proteins. Each gene consists of several thousand code words.

Germ cells come as female egg cells or male sperm cells.

Hemoglobin a protein that is capable of carrying oxygen. It is contained in red blood cells.

Hemophilia a hereditary disease. Hemophiliacs are missing the right proteins that would help stop blood from flowing in case of an injury.

Hereditary diseases illnesses that are the result of faulty genes inherited from the parents. Hemophilia is one example.

Hormones signal molecules that make cells work in a specific way.

Identical twins have exactly the same genes. They are clones.

Melanin a color substance (or pigment) made by skin cells.

Messenger RNA takes a copy of the gene (of the protein recipe) to the ribosomes where proteins are made.

Mutation a change in the genes caused by a genetic spelling error. Most mutated genes are faulty, but some are interesting and helpful. Mutations passed along to us have created all the variety we can see in the people around us.

Nucleotides molecules that build the rungs of the twisted DNA ladder.

Nucleus the ball in the middle of the cell in which the genes are stored.

Organelles the parts floating inside a cell, such as the cell nucleus, mitochondria, cell skeleton, ribosomes, and lysosomes.

Organism a complete living thing. It can consist of only one cell, such as a bacterium, or of many billions of cells, such as plants, animals, and humans.

Proteins the building blocks of cells and the tools that the cells use for their many kinds of activities. Proteins make cells, and many cells build an organism. There are 100,000 different kinds of proteins working together in the human body. Proteins are so small that a million of them would fit into less than one tenth of an inch.

Recessive genes versions of a gene that are not used for making proteins.

Red blood cells carry oxygen to all parts of the body.

Ribosomes tiny balls within a cell that make fresh proteins.

Skin cells form layers that cover the inside and outside of our body.

Sperm cells contain only 23 single chromosomes. Male sperm cells can fuse with a female egg cell in order to fertilize it.

White blood cells come in many different kinds. They are constantly on the outlook for harmful germs that they destroy.

X chromosomes chromosomes that appear as X shapes during cell division when the chromosomes are pulled apart. All 46 chromosomes in the cells of a girl are X chromosomes.

Y chromosomes tiny chromosomes that somehow appear as upside-down Y shapes. Boys have in their cells 45 X chromosomes and one Y chromosome.

Index

Alleles, 32
Antibodies, 20, 21

Bacteria, 21, 28
Bones, 12, 21
Boy, 26, 31, 37
Brain, 40

Cartilage, 12
Cell division, 16, 28
Cell tissue, 11, 22, 26
Cells, 8, 10, 11, 13, 20
Chromosomes, 11, 14, 15, 24, 29, 30, 31, 36, 37
Color blindness, 37

Daughter cells, 25
DNA, 10, 11, 16, 18, 38

Egg cell, 20, 24, 25, 26, 28, 29, 30
Embryo, 26
Endoplasmic reticulum, 12
Enzymes, 21

Father, 24, 26, 28, 29
Faulty genes, 32, 36, 38
Fertilization, 24, 25
Food, 12

Genes, 8, 12, 16, 18, 19, 20, 22, 24, 25, 32, 34, 40
Germ cells, 30
Girl, 26, 31, 37
Golgi apparatus, 12

Hair color, 32, 34
Hemoglobin, 20, 38
Hemophilia, 36, 37
Hereditary diseases, 32, 36, 38
Hormones, 21

Lysosomes, 12

Malaria, 38
Melanin, 38
Messenger RNA, 18
Mitochondrion, 12
Mother, 24, 28, 29
Mucous mebrane, 12
Muscles, 12, 21
Mutation, 38, 39

Nerve cells, 21
Nucleotides, 16, 18, 19
Nucleus, 12, 14

Organelles (cell parts), 11, 12
Organism, 21, 25

Parents, 30, 34
Personality, 40
Proteins, 11, 12, 16, 18, 19, 20, 21, 22, 26

Recessive genes, 32
Red blood cells, 20, 38
Ribosome, 12, 16

Sickle cells, 38
Skin cells, 21, 22, 23, 28, 38
Sperm cells, 24, 26, 28, 29

Twins, 8, 10, 34, 40

Viruses, 21

White blood cells, 20

X chromosomes, 14, 26, 36, 37

Y chromosomes, 14, 26, 36, 37